# The Maryland Criminal Justice System
## Super Supplement

PATRICK J. O'GUINN, SR., JD, MPA
PROFESSOR CRIMINAL JUSTICE & CO-DIRECTOR COMPUTER FORENSICS
HOWARD COMMUNITY COLLEGE
COLUMBIA, MARYLAND

THOMSON

Australia · Canada · Mexico · Singapore · Spain · United Kingdom · United States

# The Maryland Criminal Justice System
## PATRICK J. O'GUINN, SR., JD, MPA

Executive Editors:
Michele Baird, Maureen Staudt &
Michael Stranz

Project Development Manager:
Linda deStefano

Sr. Marketing Coordinators:
Lindsay Annett and Sara Mercurio

Production/Manufacturing Manager:
Donna M. Brown

Production Editorial Manager:
Dan Plofchan

Pre-Media Services Supervisor:
Becki Walker

Rights and Permissions Specialist:
Kalina Ingham Hintz

Cover Image
Getty Images*

The Adaptable Courseware Program consists of products and additions to existing Thomson products that are produced from camera-ready copy. Peer review, class testing, and accuracy are primarily the responsibility of the author(s).

ISBN: 978-1-4266-3726-1
ISBN: 1-4266-3726-8

### International Divisions List

Asia (Including India):
Thomson Learning
(a division of Thomson Asia Pte Ltd)
5 Shenton Way #01-01
UIC Building
Singapore 068808
Tel: (65) 6410-1200
Fax: (65) 6410-1208

Australia/New Zealand:
Thomson Learning Australia
102 Dodds Street
Southbank, Victoria 3006
Australia

Latin America:
Thomson Learning
Seneca 53
Colonia Polano
11560 Mexico, D.F., Mexico
Tel (525) 281-2906
Fax (525) 281-2656

Canada:
Thomson Nelson
1120 Birchmount Road
Toronto, Ontario
Canada M1K 5G4
Tel (416) 752-9100
Fax (416) 752-8102

UK/Europe/Middle East/Africa:
Thomson Learning
High Holborn House
50-51 Bedford Row
London, WC1R 4LS
United Kingdom
Tel 44 (020) 7067-2500
Fax 44 (020) 7067-2600

Spain (Includes Portugal):
Thomson Paraninfo
Calle Magallanes 25
28015 Madrid
España
Tel 34 (0)91 446-3350
Fax 34 (0)91 445-6218

# *The Maryland Criminal Justice System*
## Super Supplement
## Contents

# ABOUT THE AUTHOR

Patrick J. O'Guinn, Sr., JD, MPA, is a Professor of Criminal Justice, Co-Director of Computer Forensics and the Criminal Justice Coordinator at Howard Community College in Columbia, Maryland. Professor O'Guinn's extensive background as a criminal justice educator, trial lawyer, computer forensics examiner and former Silicon Valley police officer provides students with a fresh and stimulating perspective from which to view the modern criminal justice system. Professor O'Guinn was awarded the Outstanding Faculty of the Year Award for Howard Community College in 2000 and is the former college Diversity Chairperson. Recently, he received the 2007 Outstanding Service Award by the Maryland Criminal Justice Consortium of College Educators for his Contributions to Criminal Justice Higher Education. He is a Maryland Bar Fellow; a member of the Maryland and District of Columbia Bar; and has handled several complex and high profile court cases during his legal career. An active leader in his community, Professor O'Guinn has served on the Howard County Executive's Public Safety Transition Team in 1999 and again in 2007. He is currently serving on the Howard County, Maryland Judicial Nominating Committee.

# INTRODUCTION TO THE MARYLAND CRIMINAL JUSTICE SUPPLEMENT

The Maryland Introduction to Criminal Justice Supplement is designed for Introduction to Criminal Justice students and instructors in Maryland. It can easily be adapted to the Smith/Cole, Gaines/Miller, Bohm/Hailey text or to any other Introduction to Criminal Justice text because of it's focused explanations of major criminal justice concepts; use of thought - provoking Maryland crime data; Maryland gang data charts; Maryland court caseload and sentencing charts;Maryland incarceration data;computer/digital forensics coverage and coverage of the Maryland Eastern Shore criminal justice system.

More significant is the Supplement's inclusion of meaningful student course assignments that help to inspire reflective student thought while framing student research and writing projects in an easy to understand manner. Students and instructors will gain tremendous insights through participation in the group discussion question  assignments, and while using the targeted study questions for
learning, review and examination preparation.  The range of assignments contained in this Supplement will provide the criminal justice student with multi-sensory learning experiences through reading, writing, group discussions, research, conducting guided personal interviews with criminal justice professionals, and making guided observations of actual criminal court  proceedings in the local community.

A sample Introduction to Criminal Justice course description containing course objectives approved by the statewide Maryland Criminal Justice Consortium of College Educators is included to help get instructors off to a quick start, along with a first week student survey to help measure student background, interest, and preparedness in criminal justice studies. The student survey can also be used as part of any college's outcomes assessment data collection process.

The strength of this Supplement is it's simplification of the criminal justice process so that any reader can appreciate this dynamic field of study. The author uses his  extensive experience as a criminal justice educator, trial lawyer, former police officer and computer forensics examiner to give the reader a richer perspective about the course material from which students can formulate their new perspectives about the criminal justice system.

The Supplement's chapters match the content in the majority of the criminal justice texts in use today with added sections emphasizing Maryland gangs, sex offenses, and the Maryland Eastern Shore criminal justice system.

The Supplement's chapters focus entirely on the nature of the administration of criminal justice in Maryland and promote enhanced student awareness by introducing new criminal justice terminology and interesting Maryland criminal justice procedures. This Supplement provides an extensive review of Maryland assault-type crimes, drug offenses, sexual assault offenses, stalking and harassment offenses that can often occur on college campuses. By reading the main classroom text together with this Supplement, students will become more informed about the true operation of the Maryland criminal justice system and can be expected to make better decisions about their criminal justice careers.

## THE CHAPTERS

Each chapter in the Supplement flows naturally into the next chapter by building upon the general criminal justice concepts contained in each of the preceding chapters. Our step by step approach guides the student through selective course material to eventually develop an overall understanding of how the criminal justice system works together. Each chapter is divided into sections that provide an overview of important criminal justice principles; provides references to relevant state criminal statutes contained in the Annotated Code of Maryland; and may be followed by brief descriptions or edited references to Maryland court opinions and Maryland Rules that interpret the referenced statutes.

Some sections in the Supplement contain charts; samples of various Maryland criminal justice forms for students to complete online or as paper exercises; a sample court petition for record expungement, and important statistical research data to aid students in developing their own thoughts about the levels of crime in several identified Maryland jurisdictions.

A group of easily adaptable written assignments is included in the Appendix at the end of the Supplement for *student group work, library research, criminal justice interviews, court observations,* and *targeted study questions* that are designed to help students learn the core course material and prepare for examinations.

The Criminal Justice Supplement chapters are not intended to cover all of the course material to the same degree or in the same manner as your text. They are also not designed for a particular criminal justice instructor, but can be helpful for all criminal justice instructors. As a practical consideration, instructors will provide students with greater information in areas not covered in this Supplement. It is recommended that students focus on applying the topics covered in this Supplement to their everyday lives, or to the lives of their family and friends.

If this Supplement makes the study of criminal justice easier to understand, more interesting, or just a little bit better than anticipated, our goal of assisting in your criminal justice education has been accomplished.

# CHAPTER ONE

# THE MARYLAND CRIMINAL JUSTICE SYSTEM

## INTRODUCTION

This chapter introduces you to the Maryland criminal justice system in a simplified and practical manner. The goal is to provide you with a convenient way to view the functional relationships between law enforcement, the courts and the corrections system. Students will become familiar with important criminal justice terminology, sources of Maryland criminal law, the criminal justice process and will briefly explore new areas dealing with specific types of charging documents particular to the Maryland criminal justice system, and the interpretation of violent crime statistics.

## SECTION 1: MARYLAND CRIMINAL JUSTICE

Historically, Maryland has been recognized as a **common law state** because in Article 5 of the Maryland Declaration of Rights Maryland it adopted the laws that existed in England when Maryland won its independence from England on July 4th 1776. However, today Maryland can be considered as a mixed law state retaining some of its common law criminal offenses while increasingly codifying older common law crimes into its two volume criminal code known as the Annotated Code of Maryland Criminal Law Articles, and the single volume of the Annotated Code of Maryland Criminal Procedure Article.

As a general rule, most of the Maryland law that you may reasonably need to research on a particular subject has never been centrally located in a single penal code volume or publication. Therefore, the criminal justice student must be aware, cautiously seek out and research one or more sources of law covering a specific topic in order to ascertain the true implications of a particular law, case or rule at any given point in time. Typically, the Maryland statutory codes referenced below are called "articles" when discussed by attorneys and judges in criminal court proceedings and in appellate court opinions (*i.e.* Maryland Criminal Law Article § 2-463).

## Sources of Maryland State Law

The following references identify the principal sources of Maryland criminal law:

- State Constitution – Maryland Declaration of Rights
- Annotated Code of Maryland Criminal Law Article
- Annotated Code of Maryland Criminal Procedure Criminal Law Article
- Annotated Code of Maryland Courts & Judicial Proceedings Article
- State Case Law decisions
- Annotated Code of Maryland- Maryland Rules (Criminal Causes)

## Sources of Federal Constitutional Law

The following references identify the principal sources of U.S. Constitutional law that govern the overall operation of the criminal justice system in the U.S.:

- U.S. Constitution
- U.S. Supreme Court decisions
- U.S. Federal District Court decisions (appellate court opinions)
- Federal Codes and Statutes
- Administrative Agency Regulations & Rules

It is important to note that prior to the adoption of the current two volume set of the Annotated Code of Maryland Criminal Law Article and the Maryland Criminal Procedure Article in 2002, the primary source of Maryland criminal law was contained largely in the Article 27 Annotated Code of Maryland "Crimes and Punishment".

## Maryland State Constitution

The **Maryland Declaration of Rights** is the name of the **Maryland State Constitution.** Periodically the Maryland Declaration of Rights is cited by the appellate courts as an important factor in deciding the proper application of the Maryland criminal law to the cases that the court decides. However, with very few exceptions over the years, many individuals in Maryland are unlikely to be familiar with the Maryland Declaration of Rights as the state constitution.

One explanation for this lack of familiarity may be attributed to the fact that information about the Maryland state constitution is not routinely taught to students in high school as part of any required government or political science class.

Consequently, this Maryland Criminal Justice Supplement may provide the reader with a first exposure to the Maryland Declaration of Rights as an important foundational cornerstone of the Maryland criminal justice system. Hopefully, this knowledge of the Maryland constitution will prompt curious students to make further inquiry about how this important document has impacted the adjudication of criminal justice in the State of Maryland.

Of greater importance today is that you now know about the Maryland Declaration of Rights and realize that it provides important fundamental constitutional protections for Maryland residents that parallel the protections contained in the US. Constitution.

## Criminal Justice System - Society's Last Line of Defense

We may commonly think of the criminal justice system as a large *organized* association of related governmental agencies that systematically work together to accomplish a common goal of doing justice. However, the criminal justice system actually exists as a *loose* combination of governmental agencies at the federal, state and local levels consisting of the police, courts, and correctional facilities with a commonly stated goal of doing justice. In reaching the goal, each agency has vastly different methods of carrying out society's business of doing the justice that the people deserve.

At its core, the criminal justice system exists as an institution of **social control** whose primary role is to persuade people to voluntarily obey the positive established values of society, to refrain from conduct that has been classified as bad, undesirable or legally prohibited. The criminal justice system differs from other institutions in society in three distinct ways:

- It exists to dissuade people from violating society's values through conduct that leads to crime;
- It is society's last line of defense when people refuse to voluntarily
  obey the laws that promote living and working together in harmony;
- The criminal justice system can utilize the **force** and **punishment** of the law to *make people comply* with established standards of societal conduct when voluntary compliance with the law has failed.

## What Happens When A Crime Occurs in Maryland?

Generally, if a person commits a crime or is alleged to have committed a crime they may be:

**Charged:** This means that formal criminal charges have been filed in court against the suspect.

**Arrested:** This means that the person is physically seized and is being held by a lawful authority.

**Booked:** This means that a local law enforcement agency record of the arrest is made which includes the arrestee's name, fingerprints, photograph, and possibly DNA samples that are taken in felony cases. The local booking identification number stays with the arrested individual for life.

**Prosecuted:** This means that a Prosecutor (State's Attorney) looks at the case facts and decides whether to go forward with a trial of the case, or releases the defendant without further action. If the decision is made to proceed to trial, a summons may be issued, a warrant may be sought for the defendant's appearance in court for trial, or a grand jury indictment may be sought against the defendant in serious felony cases.

## Charging Documents

In order for a criminal case to begin its journey through the criminal justice system, there must be some mechanism for initiating the criminal justice process and notifying the court that criminal allegations have been filed against a defendant. The court must thereafter be able to provide adequate notice to the defendant of his constitutional right to due process under the law, a hearing and the opportunity to be heard and defend. A properly filed charging document commences the criminal justice process and provides official notice to the accused of the criminal charges against him. Maryland Rule 4-201.

A **charging document** can be represented by any of the following:

**Criminal Complaint:** In Maryland, a Criminal Complaint is usually in the form of an Application for a Statement of Charges that results in the issuance of a formal Statement of Charges that lists the actual criminal charges that have been filed against the alleged suspect. In Maryland, anyone can file a statement of charges.

However, in most cases, the Application for Statement of Charges is usually completed by the investigating law enforcement officer on behalf of the alleged victim and filed with the court. In other instances, usually when the police may not feel very confident about the strength of the probable cause evidence in a particular case, they may advise the victim of his right to make a personal appearance before the District Court Commissioner to request the issuance of a Statement of Charges against the suspect.

A actual copy of an **Application for Statement of Charges** is contained in the **Appendix** to this Supplement and can be found online at the Maryland District Court website http://www.courts.state.md.us/district/forms/criminal/dccr44.pdf Form DC/CR 44.

**Student Exercise 1 :**_____

Now take a moment to go online and use the Online Statement of Charges found at http://www.courts.state.md.us/district/forms/criminal/dccr44.pdf on the District Court website listed above to see if you can properly complete the **Application for a Statement of Charges** as a class exercise, or for a fictitious criminal offense that could possibly happen in your community such as an assault, battery, a stolen IPod, identity theft, a stolen mountain bike, vandalism, a stolen laptop computer, or the receipt of a bad check, etc.

---

**Criminal Information:** A Criminal Information outlines the charges against the accused and is filed with the court by the **Prosecutor** to <u>inform</u> the court of the criminal charges that have been filed against the Defendant. It is not a special document. A criminal information is just a white piece of paper with black lettering on it captioned "criminal information" that is filled out by the Prosecutor and contains the suspect's name, the applicable criminal charges, and is filed with the court to inform the court that the accused has been charged with a criminal offense.

What is important to consider, is that at the time that the criminal information is filed, we do not actually know whether the charges truly have any legal merit, or whether there is sufficient evidence to support the charges. The only thing that we are sure of is that the Prosecutor has decided to charge the defendant with a crime. In these instances, if the defendant is charged with a felony (a serious crime) by use of a criminal information, the defendant is entitled to have a preliminary hearing (probable cause hearing) that must be requested by the defendant within 10 days after the criminal information charges have been filed with the court.

**Indictment:** An Indictment is another white piece of paper with black lettering on it that is captioned "true bill indictment" and is issued by a Grand Jury Panel consisting of 23 people who meet in secret to determine if the law has been broken. Typically the grand jury listens to a police officer explain her investigative findings, or to prosecution witnesses in the presence of a prosecutor. The grand jury is allowed to **make their own finding** (or determination) of whether probable cause exists to issue criminal charges against the accused. Defense attorneys are not permitted to be present during the grand jury testimony or deliberations.

**Traffic Ticket or Uniform Criminal Citation:** A traffic ticket or Uniform Criminal Citation is also considered to be a charging document that informs the suspect of the charges against her, and constitutes the suspect's written promise to appear in court when, and where notified in the future. It is not an admission of guilt and is considered to be a *cite and release in lieu of arrest-type action.* Persons requested to sign a traffic citation issued after a traffic stop should always sign the citation in Maryland, irrespective of whether they disagree with the officer's allegations that they were speeding or otherwise violating the law.

A **refusal to sign the citation** will result in the officer arresting the driver, booking her, and taking her before a District Court Commissioner, who will then advise the driver of the traffic charges and release her with a court date to appear and answer on the charges. This is the same outcome that could have been accomplished if the driver had just signed the traffic citation at the time of the traffic stop.

Oftentimes, depending on the jurisdiction in Maryland, an arrestee may spend between 2-6 hours or more in jail waiting to see the District Court Commissioner, and may possibly have to wait overnight if arrested on the weekend. The reality is that the driver will never win the fight contesting a wrongfully issued citation with the police officer in the street. That's why we have a court system.

## UNIFORM CRIMINAL CITATION
### State of Maryland vs.

Defendant's (Last) Name    First    Middle

Current Address in Full

City    State    Zip Code

DOB    Height  Weight  Sex  Race  Hair    Eyes

Related Citations    Telephone No.
Day:    Night:

It is formally charged that the above named person on .......................... Year

at ........................M at ..................... (Location)
.........................................................., City/County, Maryland

did ....................................................................
...........................................................................
...........................................................................
...........................................................................
...........................................................................
...........................................................................

in violation of: ☐Md. Ann. Code ☐COMAR / Agency Code ☐Common Law of Md. ☐Ordinance
☐Public Local Law

Document/Article    Section    CJIS Code

Penalty:

TO ANSWER THE ABOVE CHARGE LODGED AGAINST YOU:
YOU ARE HEREBY SUMMONED AND COMMANDED TO APPEAR FOR TRIAL IN THE DISTRICT COURT
OF MARYLAND FOR .............................. (CITY/COUNTY) LOCATED AT ...............
...................................................................... MARYLAND
☐ ON............................................... AT .......................M.
☐ WHEN NOTIFIED BY THE COURT. Date
    YOUR FAILURE TO OBEY THIS CITATION MAY RESULT IN THE ISSUANCE OF
A WARRANT FOR YOUR ARREST.
    To request a foreign language interpreter or a reasonable accommodation under the Americans
Disabilities Act, please contact the court immediately.

I sign my name as a receipt of a copy of this citation and not as an admission of guilt. I hereby submit to the
jurisdiction of the Court and agree to appear when notified.

X Defendant's Signature
I solemnly affirm under the penalties of perjury that the contents of the foregoing citation are true to the best of
my knowledge, information, and belief.

Officer's
Signature
    Date    Agency  Sub-Agency  ID No.
DC/CR 45 (Rev. 5/2002)

---

Note to Law Enforcement: Remove this first copy of citation before
entering witness information. You may enter address of defendant
as shown on driver's license if that address is different from current
address.
TO THE DISTRICT COURT:
PLEASE SUMMONS THE FOLLOWING WITNESSES:

NAME

ADDRESS

CITY    STATE    ZIP

DAY PHONE    ROOM #
NIGHT PHONE    APT #
if Law Enforcement ☐ Agency  ☐ Sub-Agency  ☐ I.D.

NAME

ADDRESS

CITY    STATE    ZIP

DAY PHONE    ROOM #
NIGHT PHONE    APT #
if Law Enforcement ☐ Agency  ☐ Sub-Agency  ☐ I.D.

NAME

ADDRESS

CITY    STATE    ZIP

DAY PHONE    ROOM #
NIGHT PHONE    APT #
if Law Enforcement ☐ Agency  ☐ Sub-Agency  ☐ I.D.

NAME

ADDRESS

CITY    STATE    ZIP

DAY PHONE    ROOM #
NIGHT PHONE    APT #
if Law Enforcement ☐ Agency  ☐ Sub-Agency  ☐ I.D.

NAME

ADDRESS

CITY    STATE    ZIP

DAY PHONE    ROOM #
NIGHT PHONE    APT #
if Law Enforcement ☐ Agency  ☐ Sub-Agency  ☐ I.D.

NAME

ADDRESS

CITY    STATE    ZIP

DAY PHONE    ROOM #
NIGHT PHONE    APT #
if Law Enforcement ☐ Agency  ☐ Sub-Agency  ☐ I.D.

---

## Uniform Criminal Citation

## Rear View -Criminal Citation

## Maryland Uniform Citation Advisement-Right to Counsel
*Source: http://www.courts.state.md.us/district/forms/criminal/*

**Figure 1**

## BALTIMORE CITY CRIMINAL CITATION

This Citation constitutes prima facie evidence of the facts contained if trial is waived or where the presence of the issuing officer is not requested at trial. Failure to obey any requirement of the Citation could result in the issuance of an arrest warrant and increased fines and penalties.

## NOTICE TO DEFENDANT

To the Person Charged:
1. This paper charges you with committing a crime.
2. If you have been arrested, you have the right to have a judicial officer decide whether you should be released from jail until your trial.
3. You have the right to have a lawyer.
4. A lawyer can be helpful to you by:
    A. explaining the charges in this paper;
    B. telling you the possible penalties;
    C. helping you at trial;
    D. helping you protect your constitutional rights; and
    E. helping you get a fair penalty if convicted.
5. Even if you plan to plead guilty, a lawyer can be helpful.
6. If you want a lawyer but do not have the money to hire one, the Public Defender may provide a lawyer for you. The court clerk will tell you how to contact the Public Defender.
7. If you want a lawyer but you cannot get one and the Public Defender will not provide one for you, contact the court clerk as soon as possible.
8. DO NOT WAIT UNTIL THE DAY OF YOUR TRIAL TO GET A LAWYER. If you do not have a lawyer before the trial date, you may have to go to trial without one.

## NOTICE

A trial date notice will be sent to the address shown on the Citation. If you reside at an address different from that shown, or you change your address, you must immediately notify the Court of the correct address.

**Baltimore City Criminal Citation – Notice to Defendant**

**Maryland Uniform Citation Advisement-Right to Counsel**
*Source: http://www.courts.state.md.us/district/forms/criminal/*

**Figure 2**

# Maryland Criminal Offenses - Two Basic Classifications

Traditionally, criminal justice studies have largely defined a **misdemean**or as a minor crime punishable by *up to one year imprisonment*, fines, restitution and/or probation. Similarly, a **felony** has traditionally been defined as a serious crime punishable by *imprisonment of more than one year or death*. However, Maryland has never followed this traditional punishment – oriented crime classification framework to help us distinguish whether we are dealing with a misdemeanor or a felony offense.

Consequently, in Maryland you must look closely at the statute for each separate criminal offense with which an individual is charged to determine whether you are in fact dealing with a misdemeanor or a felony offense. This is partly because in Maryland there are misdemeanors for which you can get felony level time if we were to comparatively base our analysis of crime on the traditional criminal justice text book definitions of misdemeanors and felonies. Stated differently, there are many misdemeanor offenses in Maryland for which you can receive punishments that exceed a one year term of incarceration.

**The Maryland crime classification definitions are as follows:**

**Misdemeanor:** A misdemeanor crime in Maryland retains its original common law misdemeanor definition, or exists as proscribed by statute, unless otherwise codified by recent changes in the criminal code as many former common law misdemeanors have been in Maryland.

**Felony:** A felony crime in Maryland retains its original common law felony definition, or exists as proscribed by statute, unless otherwise codified by recent changes in the criminal code as many former common law felonies have been in Maryland.

In the recent past, a person convicted of a common law offense in Maryland could receive any sentence from the court that was not **cruel and unusual**. Consequently, prior to codification of the crime of assault, the courts had concluded that a sentence of 20 years was not cruel and unusual.

It is important to keep these distinctive Maryland misdemeanor and felony definitions in mind because a person involved in a simple fist fight with his neighbor today, would commit a Maryland misdemeanor that would result in a charge of battery or second degree assault and could receive up to 10 years of incarceration. Whereas in other states, such as California, where this conduct would constitute a misdemeanor crime of assault/battery under California Penal Code §240/242 with a maximum penalty of 6 months in the local county jail. Another common example of this distinction in Maryland is the charge of possession (of any amount) of a controlled dangerous substance other than marijuana, a misdemeanor punishable by up to four years of incarceration.

In Maryland, everyday **traffic violations** such as speeding or running a stop sign that are punishable only by a fine, are also classified as **misdemeanor offenses**, not infractions as is the case in some other states. Even vehicle equipment or repair violations known as "fix it" tickets in some states, are considered misdemeanors in Maryland. However, these minor traffic offenses, although misdemeanors, are not really viewed as true criminal offenses because they carry no potential for incarceration time. Consequently, these fine-only misdemeanor offenses are often excluded from disclosure on most job application questionnaires. By contrast, there are true criminal - traffic offenses in Maryland such as drunk driving, reckless driving, driving on a suspended or revoked license, or vehicular manslaughter that can result in a criminal conviction for which a driver can receive substantial jail time.

## Crime in Maryland

The level of crime in Maryland, as in other states, has been an increasing public concern during the past 20 years, as reflected by the number of new laws that have been passed each year to deal with modern issues related to drug offenses, murder, handgun violations, child abuse, sexual offenses, domestic violence, carjacking, computer crimes, identity theft, home invasions, witness intimidation, terrorism, assault on police officers and gang activity.

One jurisdiction, **Baltimore City**, is frequently illustrated in television and print media as having experienced more significant types of violent criminal activity than other surrounding counties in the State of Maryland. This focus may be due in part to Baltimore's size and location within the Baltimore-Washington corridor, it's geographic proximity to other metropolitan areas such as Washington, D.C., and the easy accessibility to interstate I-95 traffic, a popular interstate drug courier route on the East Coast. Moreover, Baltimore has had the unfortunate honor of having one of the highest murder rates in the nation for more than sixteen years, and at times being interchangeably considered with other states for the title of the murder capital of the nation. In **2003** according to a statistical murder rate comparison of Uniform Crime Statistics for large cities contained at www.geocites.com, **Baltimore** ranked **fourth** in the nation in the number of **murders** behind Detroit, Washington, D.C. and New Orleans. Although the national murder rankings have varied since 2003, Baltimore continues to lead the State of Maryland in murders in 2007.

The following crime statistics are taken from the Maryland Uniform Crime Reports (UCR) as reported to the FBI, and reflect only the **reported crimes** that have been made known to law enforcement officials. Certainly, reported crimes do not give us the entire picture about the actual level of crime in Maryland. However, they do give us a concrete starting point for analysis and serve as an initial crime barometer from which we can draw preliminary conclusions about the current level of criminal activity in any Maryland jurisdiction.

We should keep in mind that the annual submission of crime activity statistics by each law enforcement agency to the FBI is voluntary, may be subject to each agency's own biases in interpreting and classifying their own

| A. Statistics by Type of Crime: Totals and Percent Change From Previous Year |
| --- |

departmental activity as either felony or misdemeanor incidents according to varying criteria, and that all law enforcement agencies do not routinely participate in the UCR reporting process.

Let's look at a sampling of the crime statistics in Baltimore City, Maryland to see if there are any identifiable patterns of criminal activity about which we should be concerned, or from which we can make some preliminary conclusions.

## Crime Statistics: Baltimore City

| Year | | Rape | Robbery | Aggravated Assault | Breaking and Enter | Larceny Theft | Motor Veh Theft | Grand Totals | Change from Previous Year |
|------|-----|------|---------|--------------------|--------------------|---------------|-----------------|--------------|---------------------------|
| 1985 | 213 | 594  | 7,794   | 6,949              | 13,988             | 31,810        | 6,027           | 67,375       |          |
| 1986 | 240 | 660  | 8,008   | 6,369              | 14,388             | 30,796        | 6,884           | 67,345       | 0.00%    |
| 1987 | 226 | 595  | 7,485   | 6,030              | 13,558             | 31,270        | 7,495           | 66,659       | -1.00%   |
| 1988 | 237 | 519  | 7,418   | 6,592              | 14,359             | 33,531        | 8,468           | 71,124       | 6.70%    |
| 1989 | 259 | 543  | 7,986   | 6,876              | 14,446             | 34,082        | 8,188           | 72,380       | 1.80%    |
| 1990 | 305 | 691  | 9,491   | 7,519              | 14,867             | 36,333        | 9,939           | 79,145       | 9.30%    |
| 1991 | 304 | 702  | 10,793  | 7,295              | 16,394             | 40,406        | 10,618          | 86,512       | 9.30%    |
| 1992 | 335 | 754  | 12,290  | 8,481              | 16,503             | 41,836        | 11,332          | 91,531       | 5.80%    |
| 1993 | 353 | 668  | 12,408  | 8,577              | 18,076             | 42,814        | 10,672          | 93,568       | 2.20%    |
| 1994 | 321 | 639  | 11,303  | 8,748              | 16,026             | 43,636        | 13,603          | 94,276       | 0.80%    |
| 1995 | 325 | 684  | 11,397  | 9,172              | 16,705             | 46,750        | 11,210          | 96,243       | 2.10%    |
| 1996 | 333 | 643  | 10,429  | 8,216              | 14,887             | 43,177        | 11,186          | 88,871       | -7.70%   |
| 1997 | 313 | 480  | 8,665   | 8,072              | 12,841             | 39,581        | 8,856           | 78,808       | -11.30%  |
| 1998 | 315 | 470  | 7,718   | 7,605              | 13,279             | 36,853        | 7,375           | 73,615       | -6.60%   |
| 1999 | 305 | 374  | 7,462   | 10,536             | 12,386             | 37,524        | 7,255           | 75,842       | 3.00%    |
| 2000 | 261 | 366  | 6,634   | 8,774              | 10,751             | 32,134        | 7,871           | 66,791       | -11.90%  |
| 2001 | 256 | 299  | 5,762   | 8,520              | 10,960             | 30,457        | 8,199           | 64,453       | -3.50%   |
| 2002 | 253 | 179  | 4,764   | 8,667              | 8,814              | 27,301        | 6,572           | 56,550       | -12.30%  |
| 2003 | 270 | 208  | 4,364   | 6,385              | 7,855              | 23,307        | 6,874           | 49,263       | -12.90%  |
| 2004 | 276 | 182  | 4,085   | 7,199              | 8,022              | 21,819        | 6,731           | 48,314       | -1.90%   |
| 2005 | 269 | 162  | 3,935   | 6,943              | 7,388              | 20,132        | 6,232           | 45,061       | -6.70%   |
| 2006 | 276 | 139  | 4,260   | 6,196              | 7,664              | 18,846        | 6,276           | 43,657       | -3.10%   |

Source: Governor's Office of Crime Control & Prevention, http://www.goccp.org/four/home.php

**Figure 3**

# Crime Statistics: Baltimore City

**B. Baltimore City-Statistics by Category:  Violent Crime and Property Crime***

| YEAR | Violent Crime | Property Crime | Violent Crime Percentage | Property Crime Percentage | Change from Previous Year: Violent Crime | Change from Previous Year: Property Crime |
|---|---|---|---|---|---|---|
| 1985 | 15,550 | 51,825 | 23.1% | 76.9% | | |
| 1986 | 15,277 | 52,068 | 22.7% | 77.3% | -1.8% | 0.5% |
| 1987 | 14,336 | 52,323 | 21.5% | 78.5% | -6.2% | 0.5% |
| 1988 | 14,766 | 56,358 | 20.8% | 79.2% | 3.0% | 7.7% |
| 1989 | 15,664 | 56,716 | 21.6% | 78.4% | 6.1% | 0.6% |
| 1990 | 18,006 | 61,139 | 22.8% | 77.2% | 15.0% | 7.8% |
| 1991 | 19,094 | 67,418 | 22.1% | 77.9% | 6.0% | 10.3% |
| 1992 | 21,860 | 69,671 | 23.9% | 76.1% | 14.5% | 3.3% |
| 1993 | 22,006 | 71,562 | 23.5% | 76.5% | 0.7% | 2.7% |
| 1994 | 21,011 | 73,265 | 22.3% | 77.7% | -4.5% | 2.4% |
| 1995 | 21,578 | 74,665 | 22.4% | 77.6% | 2.7% | 1.9% |
| 1996 | 19,621 | 69,250 | 22.1% | 77.9% | -9.1% | -7.3% |
| 1997 | 17,530 | 61,278 | 22.2% | 77.8% | -10.7% | -11.5% |
| 1998 | 16,108 | 57,507 | 21.9% | 78.1% | -8.1% | -6.2% |
| 1999 | 18,677 | 57,165 | 24.6% | 75.4% | 15.9% | -0.6% |
| 2000 | 16,035 | 50,756 | 24.0% | 76.0% | -14.1% | -11.2% |
| 2001 | 14,837 | 49,616 | 23.0% | 77.0% | -7.5% | -2.2% |
| 2002 | 13,863 | 42,687 | 24.5% | 75.5% | -6.6% | -14.0% |
| 2003 | 11,227 | 38,036 | 22.8% | 77.2% | -19.0% | -10.9% |
| 2004 | 11,742 | 36,572 | 24.3% | 75.7% | 4.6% | -3.8% |
| 2005 | 11,309 | 33,752 | 25.1% | 74.9% | -3.7% | -7.7% |
| 2006 | 10,871 | 32,786 | 24.9% | 75.1% | -3.9% | -2.9% |

**Violent Crime = murder, rape, robbery & aggravated assault [combined]**

Source: Governor's Office of Crime Control & Prevention, http://www.goccp.org/four/home.php

**Figure 4**

## Murder Only/Crime Statistics: Baltimore City

| Baltimore City Statistics by Type of Crime: **Murder Only Totals** and **Percent Change From Previous Year** | | | |
|---|---|---|---|
| **YEAR** | **Murder** | **# Increase or Decrease** | **% Change from Previous Year** |
| 1985 | 213 | | |
| 1986 | 240 | 27 | 13% |
| 1987 | 226 | 14 | -6% |
| 1988 | 237 | 11 | 04% |
| 1989 | 259 | 22 | 09% |
| 1990 | 305 | 46 | 18% |
| 1991 | 304 | 1 | 0% |
| 1992 | 335 | 31 | 10% |
| 1993 | 353 | 18 | 5% |
| 1994 | 321 | 32 | -9% |
| 1995 | 325 | 4 | 1% |
| 1996 | 333 | 8 | 2% |
| 1997 | 313 | 20 | -6% |
| 1998 | 315 | 2 | 0% |
| 1999 | 305 | 10 | -3% |
| 2000 | 261 | 44 | -14% |
| 2001 | 256 | 5 | -2% |
| 2002 | 253 | 3 | -1% |
| 2003 | 270 | 17 | 7% |
| 2004 | 276 | 6 | 2% |
| 2005 | 269 | 7 | -3% |
| 2006 | 276 | 7 | 3% |
| | | shaded = increases | shaded = decreases |

Source: Governor's Office of Crime Control & Prevention,
http://www.goccp.org/four/home.php

Figure 5

An analysis of the violent crime statistics in Baltimore, Maryland is quite revealing  when compared to the uninformed and sensationalized exposure to crime that many of us receive from the local television news media. Using the actual Baltimore City crime statistics in this Supplement will help the student to compare and contrast the actual rate of crime in Baltimore City with the rates of crime in other Maryland communities.

A preliminary review of the Baltimore Crime **Table A** and **B** above reveal that the overall violent crime rate in Baltimore City has actually decreased each year for the past 10 years, as has been the case across most of the nation according to the FBI Uniform Crime Reports.

However, if crime has been going down in recent years, why do many of us still have more fear each year about being victimized in our cities, urban areas and suburban communities? Think for a moment about your own personal fears of being victimized and examine why those fears might still be lingering in the back of your mind today?

## Baltimore Murder Data Trend Analysis

A closer analysis of the Baltimore City violent crime statistics actually reveals a structurally inherent problem that results from the combining of multiple categories of violent crimes such as murder, rape & robbery into a single Uniform Crime Report (UCR) "violent crime" category and thereafter equally disturbing the violent crime trend data across all reported years. The net result of this aggregate data analysis is a **dilution** of the actual **murder rate** statistics for Baltimore City.

### A Closer Analysis

For example, when the murder rates are separated out from the collective violent crime data and analyzed for each of the past ten years as reflected in **Table C** above, the violent crime analysis results are dramatically higher from year to year. Murders have increased 2.6 % in 2006 over 2005, decreased 2.6 % in 2005 over 2004, and increased 2.2 % in 2004 over the 2003 murder rate. In 2003 the murder rate increased 6.3% over 2002, and decreased 12.3% in 2002 over the 2001 rate. In 2001 the murder rate decreased 3.5% over the 2000 murder rate with a decrease of 11.3% over the 1999 murder rate.

The worst murder rate in Baltimore City during the past 14 years occurred in **1993** with the occurrence of **353 killings**. Certainly, improvements in the murder rate have been made over time. However, in the final analysis, the 12 instances of increases in the Baltimore City murder rate outnumber the 8 instances of decreases in the Baltimore City murder rate during the past 19 years.

In 2007 the Baltimore City murder rate was consistently viewed by community commentators as reaching a "crisis level" resulting from repeated daily news barrages
illuminating numerous shooting victims. In a general sense, families in Maryland and throughout the U.S. frequently move or relocate to certain communities based solely upon their perception of the level of crime and the quality of the school system in the

local community

While we cannot say with absolute certainty that the perceived crime rate and school system performance in a particular community gives us the total measure of the quality of life in a community, it does become an important consideration to satisfy the inherent sense of safety and security that many residents seek today. While real estate agents have known this fact for a long time, they are prevented by law from commenting about the crime rate and school systems in a particular community because of  past racially discriminatory practices known as "redlining" that  unfairly targeted areas where decent, hard working and caring people of all races and cultural backgrounds have lived in harmony for decades.

In refocusing our attention on the overall statistical decrease in crime in many Maryland jurisdictions that we can compare with the Baltimore City crime rates, the greater questions presented today may be as follows:

- How much crime are we willing to tolerate as being acceptable in our communities in any given crime category?
- Is six homicides in a year just six killings too many?
- Do you know what level of crime actually exists in your community today?
- Are you concerned about crime or about being victimized?
- What can we really do about crime in our community?
- Who should really be concerned about controlling crime in our communities today? The police? The residents? Our government officials?

_____

**Student Exercise 2 :**_____

You can review and compare the applicable crime statistics in your own community using the Maryland Uniform Crime Reports (UCR) as reported to the FBI, and decide the best manner in which crime can be effectively controlled to your satisfaction. The questions presented above can serve as a mental checklist for use in your inquiry and analysis. Go to the Governor's Office of Crime Control & Prevention, website at http://www.goccp.org/four/home.php and evaluate your local community crime statistics. Record your overall findings in a separate report.

_____

For comparison, let's look at the murder rate statistics in other Maryland counties located within a 30-50 mile radius of Baltimore City, Maryland.

## Maryland Counties Murder Trend Data - Comparisons

**D. Maryland Statistics by Type of Crime:**
   Murder Only County Totals and Percentage Change From Previous Year

**Montgomery County**, Maryland

| YEAR | Murder | # Increase or Decrease | % Change from Previous Year |
|---|---|---|---|
| 1985 | 12 | | |
| 1986 | 8 | 4 | -30% |
| 1987 | 17 | 9 | 52% |
| 1988 | 19 | 2 | 11% |
| 1989 | 21 | 2 | 10% |
| 1990 | 25 | 4 | 19% |
| 1991 | 26 | 1 | 04% |
| 1992 | 21 | 5 | -19% |
| 1993 | 30 | 9 | 42% |
| 1994 | 34 | 4 | 13% |
| 1995 | 21 | 13 | -38% |
| 1996 | 13 | 12 | -57% |
| 1997 | 23 | 10 | 76% |
| 1998 | 13 | 10 | -43% |
| 1999 | 13 | 0 | 0% |
| 2000 | 12 | 1 | -07% |
| 2001 | 19 | 2 | 16% |
| 2002 | 32 | 13 | 68% |
| 2003 | 23 | 11 | -34% |
| 2004 | 18 | 5 | -21% |
| 2005 | 21 | 3 | 16% |
| 2006 | 19 | 2 | -10% |

shaded = increases        shaded = decreases

**Prince Georges County**, Maryland

| YEAR | Murder | # Increase or Decrease | % Change from Previous Year |
|---|---|---|---|
| 1985 | 47 | | |
| 1986 | 52 | 5 | 10% |
| 1987 | 94 | 42 | 80% |
| 1988 | 101 | 7 | 7% |
| 1989 | 122 | 21 | 20% |
| 1990 | 112 | 10 | -8% |
| 1991 | 147 | 35 | 31% |
| 1992 | 134 | 13 | -8% |
| 1993 | 141 | 7 | 5% |
| 1994 | 127 | 14 | -9% |
| 1995 | 137 | 10 | 7% |
| 1996 | 142 | 5 | 03% |
| 1997 | 83 | 59 | -41% |
| 1998 | 107 | 24 | 29% |
| 1999 | 95 | 12 | -11% |
| 2000 | 72 | 23 | -24% |
| 2001 | 109 | 37 | 51% |
| 2002 | 141 | 32 | 29% |
| 2003 | 135 | 6 | -04% |
| 2004 | 146 | 11 | 08% |
| 2005 | 164 | 18 | 12% |
| 2006 | 130 | 34 | -20% |

shaded = increases        shaded = decreases

Source: Governor's Office of Crime Control & Prevention, http://www.goccp.org/four/home.php

Figure 6

## Maryland Counties Murder Trend Data – **Comparisons (cont.)**

### D. Maryland Statistics by Type of Crime:
Murder Only County Totals and Percentage Change From Previous Year

**Baltimore County**, Maryland

| YEAR | Murder | # Increase or Decrease | % Change from Previous Year |
|------|--------|------------------------|------------------------------|
| 1985 | 24 | | |
| 1986 | 31 | 7 | 3% |
| 1987 | 32 | 1 | 03% |
| 1988 | 24 | 8 | -3% |
| 1989 | 33 | 9 | 4% |
| 1990 | 34 | 1 | 03% |
| 1991 | 25 | 9 | -3% |
| 1992 | 43 | 18 | 72% |
| 1993 | 36 | 7 | -16% |
| 1994 | 31 | 5 | -14% |
| 1995 | 38 | 7 | 23% |
| 1996 | 34 | 4 | -11% |
| 1997 | 23 | 11 | -32% |
| 1998 | 20 | 3 | -13% |
| 1999 | 30 | 10 | 50% |
| 2000 | 33 | 3 | 10% |
| 2001 | 31 | 2 | -06% |
| 2002 | 29 | 2 | -06% |
| 2003 | 31 | 2 | 06% |
| 2004 | 29 | 2 | -06% |
| 2005 | 40 | 11 | 38% |
| 2006 | 35 | 5 | 13% |
| | | shaded = increases | shaded = decreases |

**Wicomico County**, Maryland

| YEAR | Murder | # Increase or Decrease | % Change from Previous Year |
|------|--------|------------------------|------------------------------|
| 1985 | 5 | | |
| 1986 | 10 | 5 | 100% |
| 1987 | 3 | 7 | -70% |
| 1988 | 2 | 2 | -70% |
| 1989 | 9 | 7 | 400% |
| 1990 | 9 | 0 | 0% |
| 1991 | 5 | 4 | -44% |
| 1992 | 8 | 3 | 60% |
| 1993 | 3 | 5 | -63% |
| 1994 | 2 | 1 | 33% |
| 1995 | 4 | 2 | 100% |
| 1996 | 3 | 1 | -25% |
| 1997 | 7 | 4 | 100% |
| 1998 | 4 | 3 | -43% |
| 1999 | 4 | 0 | 0% |
| 2000 | 4 | 0 | 0% |
| 2001 | 3 | 1 | -25% |
| 2002 | 5 | 2 | 40% |
| 2003 | 1 | 4 | -80% |
| 2004 | 5 | 4 | 400% |
| 2005 | 4 | 1 | -20% |
| 2006 | 8 | 4 | 100% |
| | | shaded = increases | shaded = decreases |

Source: Governor's Office of Crime Control & Prevention, http://www.goccp.org/four/home.php

Figure 7

# Baltimore City - 25 Year [ Month by Month] Homicide Data

## E. Statistics by Type of Crime: Homicide by Month - Baltimore City

HOMICIDES BY MONTH: 1970 TO 2005 FROM OFFICIAL UCR DATA

| YEAR | JAN. | FEB. | MAR. | APR. | MAY | JUN. | JUL. | AUG. | SEP. | OCT. | NOV. | DEC. | TOTAL |
|------|------|------|------|------|-----|------|------|------|------|------|------|------|-------|
| 1970 | 13 | 15 | 19 | 24 | 24 | 25 | 13 | 19 | 23 | 20 | 22 | 14 | 231 |
| 1971 | 28 | 15 | 18 | 26 | 32 | 23 | 30 | 25 | 21 | 29 | 32 | 44 | 323 |
| 1972 | 32 | 22 | 25 | 23 | 19 | 25 | 35 | 45 | 30 | 22 | 26 | 26 | 330 |
| 1973 | 35 | 10 | 28 | 12 | 20 | 22 | 31 | 29 | 19 | 17 | 21 | 36 | 280 |
| 1974 | 18 | 24 | 18 | 32 | 16 | 28 | 31 | 30 | 20 | 28 | 21 | 27 | 293 |
| 1975 | 25 | 14 | 23 | 32 | 12 | 30 | 30 | 15 | 19 | 25 | 16 | 18 | 259 |
| 1976 | 21 | 20 | 18 | 14 | 12 | 18 | 25 | 15 | 10 | 13 | 19 | 15 | 200 |
| 1977 | 10 | 15 | 14 | 10 | 16 | 12 | 16 | 16 | 9 | 16 | 17 | 20 | 171 |
| 1978 | 13 | 14 | 17 | 14 | 21 | 12 | 14 | 17 | 22 | 21 | 13 | 19 | 197 |
| 1979 | 17 | 26 | 28 | 19 | 14 | 22 | 21 | 10 | 23 | 23 | 23 | 19 | 245 |
| 1980 | 15 | 19 | 16 | 15 | 20 | 18 | 22 | 24 | 22 | 13 | 17 | 15 | 216 |
| 1981 | 19 | 20 | 21 | 19 | 16 | 22 | 19 | 20 | 20 | 15 | 18 | 19 | 228 |
| 1982 | 24 | 21 | 11 | 13 | 19 | 22 | 24 | 14 | 19 | 18 | 26 | 16 | 227 |
| 1983 | 17 | 16 | 12 | 27 | 15 | 7 | 16 | 18 | 18 | 18 | 15 | 22 | 201 |
| 1984 | 21 | 22 | 17 | 19 | 19 | 17 | 12 | 14 | 21 | 23 | 21 | 9 | 215 |
| 1985 | 16 | 18 | 16 | 20 | 17 | 13 | 13 | 18 | 23 | 16 | 23 | 20 | 213 |
| 1986 | 15 | 17 | 21 | 15 | 15 | 20 | 30 | 17 | 26 | 21 | 24 | 19 | 240 |
| 1987 | 19 | 18 | 16 | 19 | 12 | 26 | 26 | 26 | 11 | 20 | 12 | 21 | 226 |
| 1988 | 21 | 14 | 25 | 24 | 12 | 11 | 20 | 21 | 18 | 19 | 23 | 26 | 234 |
| 1989 | 20 | 17 | 18 | 23 | 17 | 23 | 26 | 29 | 24 | 27 | 18 | 20 | 262 |
| 1990 | 28 | 25 | 16 | 22 | 22 | 20 | 25 | 42 | 25 | 23 | 23 | 34 | 305 |
| 1991 | 31 | 19 | 24 | 29 | 23 | 20 | 18 | 28 | 24 | 31 | 21 | 36 | 304 |
| 1992 | 29 | 26 | 20 | 22 | 25 | 28 | 37 | 29 | 34 | 27 | 26 | 32 | 335 |
| 1993 | 27 | 25 | 28 | 30 | 29 | 33 | 29 | 35 | 26 | 32 | 30 | 29 | 353 |
| 1994 | 22 | 14 | 25 | 29 | 23 | 28 | 28 | 27 | 28 | 35 | 30 | 32 | 321 |
| 1995 | 28 | 16 | 29 | 22 | 26 | 25 | 29 | 34 | 29 | 24 | 34 | 29 | 325 |
| 1996 | 27 | 16 | 27 | 36 | 32 | 30 | 31 | 39 | 20 | 30 | 19 | 24 | 331 |
| 1997 | 29 | 20 | 20 | 21 | 37 | 29 | 20 | 24 | 31 | 23 | 31 | 27 | 312 |
| 1998 | 25 | 31 | 27 | 23 | 17 | 32 | 25 | 28 | 34 | 16 | 24 | 31 | 313 |
| 1999 | 27 | 14 | 22 | 14 | 22 | 26 | 26 | 25 | 29 | 29 | 36 | 35 | 305 |
| 2000 | 24 | 16 | 26 | 33 | 22 | 20 | 28 | 25 | 16 | 19 | 15 | 17 | 261 |
| 2001 | 21 | 23 | 16 | 25 | 26 | 26 | 9 | 18 | 16 | 28 | 21 | 27 | 256 |
| 2002 | 19 | 17 | 26 | 22 | 21 | 22 | 29 | 14 | 24 | 24 | 18 | 17 | 253 |
| 2003 | 23 | 20 | 23 | 26 | 27 | 22 | 25 | 21 | 12 | 18 | 23 | 30 | 270 |
| 2004 | 23 | 22 | 21 | 18 | 28 | 25 | 31 | 23 | 27 | 23 | 21 | 14 | 276 |
| 2005 | 32 | 13 | 17 | 16 | 27 | 32 | 27 | 21 | 20 | 22 | 22 | 20 | 269 |

Source: Governor's Office of Crime Control & Prevention, http://www.goccp.org/four/home.php

Figure 8

## Student Exercise 3 :_____

Take a moment to review the **Month to Month Homicide** data in **Table E** above to see if you can identify any meaningful murder crime patterns by which, if analyzed closely, could assist law enforcement officers in their efforts to reduce the incidence of homicides occurring in Baltimore City? Record your overall findings in a separate report.

## Baltimore City *Citistat* – Crime Mapping Initiative

In an effort the reduce crime and increase government accountability, the former Baltimore City Mayor, Martin O'Malley implemented the Citistat Management Accountability Program modeled after the successful **New York City Police Department Compstat** program pioneered by Jack Maple. The Citistat program, like the New York City Compstat Program, utilizes computer mapping as a measurement tool to identify weekly problem areas in government and targets those areas for immediate improvement and results, such as the dramatic reduction in crime that was achieved in New York City in the 1990's.

The four basic principles of **Citistat** are:

- Gathering accurate and timely intelligence
- Using effective tactics and strategies to address problem areas
- Rapid deployment of governmental resources to address problem areas
- Relentless follow-up and assessment of results

A map plotting technique called **"geocoding"** is used to create a visual representation of all **Citistat** findings on a computer pin map [to identify crime patterns.

Whether or not Baltimore City's use of the Citistat program can achieve the same level of crime reduction success as New Your City remains a formidable challenge that only time will reveal.

## The Pretrial Process

In criminal cases, the suspect, once charged, is called the defendant and must appear in court for an initial appearance.

**Initial Appearance** – At the initial appearance, the defendant appears before a judge to hear the criminal charges against her and is advised by the judge of her fundamental constitutional rights, particularly the right to have a lawyer represent her at trial, and advice of her right to a preliminary if felony charges have been filed.

If the crime is a **misdemeanor** that carries less than 6 months of incarceration,
there is generally no right to a jury trial in Maryland, and the defendant will receive a summary trial before a judge in the Maryland District Court. A summary trial is also referred to as a bench trial or a court trial.

If the crime is a **felony**, there must be a probable cause showing contained in the charging document itself (i.e. some alleged degree of proof showing a minimal connection between the person charged and the commission of a felonious crime). The **bail** (an amount of money required from the defendant to get out of jail) is typically set by the court to ensure the appearance of the defendant at trial to answer the charges against him.

**Preliminary Hearing** - If the defendant is charged with a felony prior to his appearance in the District Court, he is entitled to a preliminary hearing, as long as he previously made the request for a preliminary hearing within 10 days of the date of his arrest or within 10 days of being served with a criminal court summons. Preliminary hearings are also called "probable cause" hearings and are designed to determine if there is sufficient evidence to establish that probable cause exists to believe that the defendant was in some way involved in, or connected with, the alleged felony criminal activity. The test used by the courts in preliminary hearings is whether this is "any evidence", sometimes referred to as a "scintilla" of evidence that exists to connect the defendant with the alleged crime. If so, the defendant will be held over for trial on the felony charges in the Circuit Court, if the felony charges
are not dismissed by the District Court Prosecutor.

Students should note that the preliminary hearing is not designed to determine the defendant's guilt or innocence. It is, however, the first opportunity for the defendant to test the Prosecutor's evidence before trial. In the majority of criminal cases (possibly 95%), the courts typically find that probable cause exists to hold the defendant over for trial. This is so because hearsay evidence is also admissible during the preliminary hearing and police officers usually show up and simply read their police reports or a statement of charges into the court record while testifying.

As a practical matter, many people who are charged with a felony crime often wait too long before talking with a lawyer, or just forget to request a preliminary hearing within the required 10 day period. In those instances, the right to challenge the Prosecutor's evidence for legal sufficiency in a preliminary hearing is lost or waived, and the felony charges will proceed to trial in the Circuit Court.

### "No Felony Rule" in Maryland District Courts
The Maryland District Courts are courts of **limited jurisdiction** and the general rule is that no felony cases can be tried in the Maryland District Courts. However, there are some exceptions to this rule for certain felony offenses involving theft and similar other criminal offenses under Maryland's Consolidated Theft Statute.

**Arraignment** – Is the first formal appearance of the defendant or his legal counsel in court where the criminal charges are read to the defendant in open court. At arraignment the defendant enters a plea to the charges. If the defendant refuses to enter a plea, the court will enter a "not guilty" plea on the defendant's behalf.

Even in cases where there is overwhelming evidence of the defendant's guilt (such as a confession to the crime), the only plea to be entered at the arraignment is "not guilty", unless there has been a previously negotiated plea bargain between the Prosecutor and the Defense Attorney, or the defendant is pursuing an insanity defense. This is so because there is not going to be a trial or any determination of guilt or innocence at the arraignment phase, and the practice of judges entertaining a guilty plea while handling a busy arraignment docket is uncertain and varies from jurisdiction to jurisdiction.

To save the court time, many attorneys informally advise their clients of the criminal charges against them contained in the indictment, and waive the reading of the formal charges to the defendant by the court. At that point, a copy of the indictment is given to the defendant and his attorney, and they leave the courtroom with a trial date. In many Maryland jurisdictions today, if the Defense Aounsel enters her appearance before the arraignment date, the court clerk will usually cancel the arraignment appearance, and schedule the case for trial.

**Plea Bargaining** – Approximately 90% percent or more of all cases in Maryland that are scheduled to go to trial are disposed of by plea bargaining before trial. Despite the occasional public outcries about the fairness or unfairness of the plea bargaining process (either to the victim or the defendant), plea bargains actually constitute the fuel that makes the criminal justice process work. Given the increasing number of criminal cases that are filed each year, and growing criminal court dockets, few, if any courts, are equipped to provide every criminal defendant with the opportunity to receive a jury trial.

**Insanity Plea** - Not Criminal Responsible (NCR) – In Maryland, the insanity plea is known as Not Criminally Responsible or NCR. There is no diminished capacity insanity defense in Maryland, meaning that the defendant is considered to be either 100% sane, or 100% insane. There is no in between mental status such as temporary insanity. When an insanity plea is entered, the defendant is actually admitting that he committed the crime, but at the time of the crime he was not legally responsible for his actions. That is, he was incapable of forming the required criminal intent necessary to hold him legally responsible for the crime.

An example would be a case in which a person had a sleepwalking disease known as **"automatism"** and committed a crime in his sleep without knowing it. Nevertheless, approximately,90-95% percent of all persons entering Not Criminally Responsible (NCR) pleas in Maryland are found to be responsible for their actions, and are convicted of some crime. The law related to **NCR Pleas** can be found at **Maryland Criminal Procedure Code Ann. §3-110.**

# Types of Punishment

- **Fines** – the payment of money to the court
- **Restitution** – payment to the victim for damages or loss
- **Probation** – no sentence to serve but supervision by the **court** for a stated period of time
- **Intermediate Punishment Work Release** – residence at a half way house with release to attend school or work during the day
- **Weekend Jail** – the incremental serving of time on successive weekends from Friday to Sunday
- **Commitment to Mental Health Facility** – in rare cases a defendant can be ordered committed to a hospital for mental health treatment
- **Imprisonment** – service of the sentence imposed by the court in local or state correctional facilities to include life with or without parole.
- **Death** – a sentence that is imposed by a jury after a finding of guilt and consideration of mitigating and aggravating circumstances of the defendant's life and nature of the crime

**Corrections** – Involves the administration of the punishment imposed by the court for persons who have been adjudicated guilty after trial or a plea bargain.

**Appeals** – The defendant can appeal her conviction (i.e. ask for the conviction to be overturned and a new trial granted) typically in two ways:

## a. Legal Grounds

- Based upon improper jury selection
- Based upon improper admission or exclusion of evidence at trial
- Based upon wrong interpretations of the law or instructions to the jury
- Based upon improper sentencing
- Based upon any determination that the defendant did not receive a fair trial

## b. Constitutional Grounds

- Illegal search and seizure
- Improper questioning by police
- Improper police lineup
- Lack of Speedy Trial
- Having an incompetent lawyer (i.e. in a post conviction proceeding-after all direct appeals for trial errors have been exhausted)

**Parole** – Imprisoned inmates may be eligible for parole, which is a release from state custody and supervision prior to the completion of a sentence, usually based upon the accumulation of "good time" credits for good behavior while in prison, or based upon their mandatory release date. In Maryland, inmates must service at least 50% percent of their imposed sentence for certain violent crimes before becoming eligible for release on parole. Persons released on parole are supervised by the state as compared to persons on probation who are supervised by the court.

**Bail** – Bail is the setting of a monetary payment amount that the defendant is required to pay into the court (usually in the form of a surety bond that is put up by a bail bondsman) to ensure the defendant's appearance at trial. Bail is not supposed to be used as a pretrial form of punishment to keep accused individuals in jail who have not had their day in court. However, some judges set high bails on certain defendants which in practice keeps them incarcerated, and also serves as a means to ensure that the defendant does not skip town before the trial date.

Most individuals who have never had to hire a bell bondsman become a little surprised when they learn that the **10% bail** normally paid to the bail bondsman up front on their behalf is **nonrefundable.** The bail bondsman is taking the risk that the defendant will not show up. If the defendant fails to show up for trial as scheduled, the court will call the bond in open court and give the bail bondsman (and possibly their bounty hunter friends) approximately 90 days to track down and return the bail jumping defendant to the court. Otherwise, the bail bondsman's insurance company takes a big loss on the forfeited bail bond(s) that have previously been placed with the court.

The following factors are typically considered by the court in setting a reasonable bail:
- Seriousness of the crime
- Whether the defendant poses a danger to the community
- The defendant's prior criminal history
- The defendant's ties to the community such as family, school, etc. (stability factors)
- Whether the defendant is a **flight risk** to leave the community or country before trial (i.e. has the defendant failed to appear in court before?) If yes, no bail will be granted
- The defendant's work history evidencing the overall responsibility of the client.
- For purposes of setting bail, the courts presume that the defendant
has committed the offense(s) for which he is charged, and in light of that presumption, try to set a reasonable bail that will ensure the defendant's appearance at trial

# Chapter Two

# The Maryland Court System

## MARYLAND COURTS

The Maryland Courts have a **four tier structure**. The **Court of Appeals** is the highest court in the state. The intermediate or second level court is called **the Court of Special Appeals.** The third level court, commonly known as the trial court, is the **Circuit Court**. The fourth level court, which I refer to as the neighborhood court, is the **District Court**.

It is important to know the jurisdictional authority and operational characteristics of each court within the Maryland judicial system to better understand how the judicial process affects the daily lives of Maryland residents, and the professional men and women working within the criminal justice system.

## The Maryland Court of Appeals

The **Maryland Court of Appeals** is the highest court in the State of Maryland and is located in Annapolis, Maryland, the state capital. The Court of Appeals was created in 1776 by the Maryland State Constitution. Seven judges preside over the court and one judge is appointed by the Governor from each of the seven Maryland judicial circuits. The judges can decide to hear a case that is appealed from trial by the filing of a **Writ of Certiorari,** but they have the exclusive discretion not to hear any case that does not substantially affect **important Maryland public policy or touch upon important constitutional rights.**

The Court of Appeals **hears approximately 10% percent of all cases** that are filed with the court. A Writ of Certiorari can essentially be viewed as a request by a criminal defendant to have the court "certify" her case for a hearing before the court. A significant and historical change within the Court of Appeals occurred with the appointment of **Judge Robert M. Bell as the Chief Judge of the Court of Appeals** in 1996. Chief Judge Bell became the first African American in history to lead the highest court in Maryland. Chief Judge Bell has continued to lead the court with distinction, and has been nationally recognized for his leadership in implementing statewide mediation processes in all Maryland courts.

## The Maryland Court of Special Appeals

The **Maryland Court of Special Appeals** is Maryland's intermediate court that was created in 1966. The court does not hear any *special appeals*, as the name might imply, and actually hears all appeals. The court's name is established in the Maryland Constitution, and although possibly confusing to criminal litigants who handle their own appeals (pro se), the name has endured. The Court of Special Appeals has 13 judges. In most instances a panel of three judges will hear and decide appeals of criminal convictions.

The Court of Special Appeals can be viewed as a **filtering appellate court** in the State of Maryland because it must review all criminal appeals that are filed by convicted criminal defendants after trial. However, the court is not obligated to grant a hearing to any criminal defendant who files an appeal, and most cases are routinely decided without a hearing, are decided as unreported opinions, or are denied. **Approximately 10% of the appeals** to the Court of Special Appeals are **granted a hearing.**

## The Circuit Courts

The **Maryland Circuit Courts** are the trial courts of **general jurisdiction** in the State of Maryland. Circuit courts have broad jurisdiction to hear a wide range of criminal and civil cases in each county and in Baltimore City. Generally, the most serious felony cases and some misdemeanor cases are tried in the Circuit Court. The Circuit Court is the only Maryland Court with the authority to hear and decide criminal jury trials, which consist of 12 jurors and two alternate jurors.

## The District Court

The **Maryland District Courts** are located in 34 locations throughout the state of Maryland and **approximately 90% percent of all criminal cases** that are filed, initially pass through the District Courts for bail review or trial. The District Courts are courts of **limited jurisdiction,** They have no authority to hear jury trials, and generally cannot hear felony cases with the exception of certain cases specified by law, many of which fall under Maryland's Consolidated Theft Statute. The District Court can hear and decide civil matters involving claims up to $25,000.00 dollars.

The District Court can be viewed as the **neighborhood court** because it is the principle place where the majority of the Maryland residents will appear during their lifetime to get their justice in civil, criminal, traffic, landlord-tenant, peace order, and protective order cases. The cases are heard and decided by a single judge.

## The Criminal Caseload Docket – District Court

In **January 2007** the Maryland District Courts handled approximately **19,403 criminal case dispositions**. By comparison in **January 2006,** the Maryland District Courts handled **19,095 criminal case dispositions**, an 8% increase over the 17, 681 criminal cases handled by the Maryland District Courts in January 2005. The number of cases in each court has risen steadily during the past few years as the Maryland population continues to increase. In reviewing the overall District Court caseload statistics we must ask the following questions:

- What explanations can be given for the increase in the District Court caseload?
- Can we attribute increases to improved detection by law enforcement efforts?
- Are increases the result of the enactment of new criminal laws each year?
- Are increases due to a change in the makeup of the state's population by age, ethnicity or gender?

Maryland is a diverse state. Is there any reliable correlation between the ethnic and racial make up of the population and increase in crime rates? The following Table gives us a breakdown of the ethnic and racial population in Maryland according to the U.S. Census:

## Maryland Race and Ethnicity Data

| Race/Ethnicity | Percentage of Population |
|---|---|
| White | 64% |
| African American | 29.3% |
| Latino\Hispanic | 5.7% |
| Asian | 4.8% |
| Native American Indian | .3% |
| Bi-racial | 1.5% |

**Source: 2005 Quick Facts U.S. Census Bureau**

**Figure 9**

## Maryland Drug Courts

The Maryland Drug Courts were established in 1995 to provide drug offenders with treatment as an alternative to incarceration. Each year the drug courts experience a high level of success with the program participants, and the drug concept is growing and continuing to expand within the State of Maryland as a useful community resource.

## The Maryland Court System

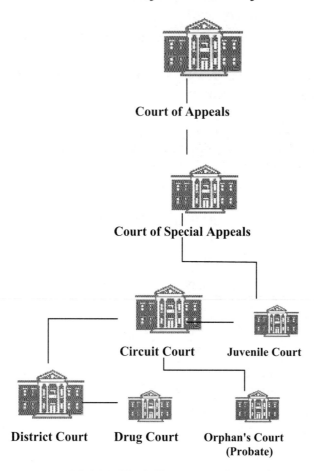

Figure 10

# CHAPTER THREE

# THE POLICE

## SECTION 3: THE POLICE

### Law Enforcement in Maryland

Police officers play a vital role in preserving the quality of life in Maryland communities. They are often called upon to act in several different capacities during the course of their daily activities while protecting the public and maintaining order.

Consequently, the public's expectations of the results that the police are able to achieve in handling a single call or incident may conflict with the many roles that police officers must play throughout a typical day such as peace keepers, law enforcers, counselors, social workers, investigators, mediators and general problem solvers. Nevertheless, a career in law enforcement provides a tremendous degree of variety for officers that can be rewarding, exciting, stressful, boring and challenging all in a single day.

### Arrest Defined

**Arrest:** In Maryland an arrest is generally defined as the substantial interference with, or substantially depriving a person of their freedom of movement such that the person reasonably believes under the circumstances at that time that they are not free to leave from the control of the police.

In many criminal justice texts, an arrest is usually defined as the taking of a person into physical custody in the manner prescribed by law. However, it is important to note that in Maryland an arrest can actually take place without the necessity of
placing handcuffs on the suspect.

_____

**Student Exercise 4:**_____
What would you think if you woke up in your home at 4:00 a.m only to find that your bed was surrounded by 7 uniformed police officers?

Would you feel sufficiently free to just walk past the several officers to go downstairs to the first floor bathroom and return to the presence of the officers when you get ready? Record your reactions and thoughts on this scenario below.

Although it may be theoretically possible for you to leave the officers' presence without permission in this scenario, it may be more likely that an individual would reasonably believe that he was not free to leave the presence of the officers at his own will without permission. Consequently, the officers' presence under these circumstances may constitute an arrest.

**Citizens Arrest:** In Maryland, there is no clear general common law right of a citizen to make a citizens' arrest. However, in Maryland, a **private citizen** can make a citizens' **arrest** when a **felony** *is in fact* committed in the **presence** of the private citizen. But why would anyone want to do so? The commission of a felony could be dangerous to a person's health and well-being while trying to make a citizens arrest. Given the level of violent crime in some areas today, the suspect may well be armed. So why not just call the police for assistance?

In **misdemeanor cases**, a private citizen can make a citizens' arrest for actions involving a **"breach of the peace"** that are committed in the citizen's **presence.** However, if a private citizen is mistaken about the law or the facts that exist at the time of a citizen's arrest, then he could be subject to potentially serious criminal charges of assault, battery, false imprisonment and even kidnapping. In general, Maryland courts tend to frown upon and discourage self help through the use of citizens' arrests, and the courts prefer private citizens to call the police for assistance, instead of taking the law into their own hands.

**Maryland Merchant's Privilege:** Since there is no general power or authority that exists for a private citizen to make a citizens arrest for a misdemeanor that does not involve a breach of the peace, then how do private store security personnel or private store detectives have the authority to detain and arrest people suspected of shoplifting?

The basic answer is that shoplifting does not usually involve a breach of the peace and therefore store detectives would have difficulty justifying their actions in making a private citizens arrest in Maryland if there was no statutory right to make an arrest under the **Maryland Merchant's Privilege law**, or the common law **shopkeepers** right to prevent the theft of his property.

Under the **Merchants' Privilege Law contained** in Courts and Judicial Proceedings Article § 5-402, store detectives have the authority to make arrests of suspected shoplifters and are **immune from all civil liability**, provided that they act reasonably under such circumstances, such that a reasonable merchant, store detective or shopkeeper would believe that she had witnessed a shoplifting incident occur in her presence.

# Maryland Merchants Privilege Arrest Statute

## Md. COURTS AND JUDICIAL PROCEEDINGS Code Ann. § 5-402 (2007)

§ 5-402. Civil liability of merchant or owner or lessee of motion picture theater for
detention

(a) Merchants. -- A merchant or an agent or employee of the merchant who detains or causes the arrest of any person shall not be held civilly liable for detention, slander, malicious prosecution, false imprisonment, or false arrest of the person detained or arrested, whether the detention or arrest takes place by the merchant or by his agent or employee, if in detaining or in causing the arrest of the person, the merchant or the agent or employee of the merchant had, at the time of the detention or arrest, probable cause to believe that the person committed the crime of "theft," as prohibited by *§ 7-104 of the Criminal Law Article*, of property of the merchant from the premises of the *merchant*.

(b) Motion picture theater owner or lessees. -- An owner or lessee of a motion picture theater or an agent or employee of the owner or lessee who detains or causes the arrest of any person may not be held civilly liable for detention, defamation, malicious prosecution, false imprisonment, or false arrest of the person detained or arrested, if in detaining or causing the arrest of the person, the owner, lessee, agent, or employee had, at the time of the detention or arrest, probable cause to believe that the person committed in the motion picture theater a violation of § 7-308(e) of the Criminal Law Article.

## Probable Cause – Defined

**Probable Cause** [to make an arrest].  Probable cause is defined as:

- that degree of evidence or information **known to the police** officer
- **at the time** and under the circumstances
- that **reasonably leads her to believe** that a crime has been committed
- and the suspect is the person responsible for committing the alleged crime

A police officer does not have to have absolute proof that a crime has been committed in order to make an arrest.  Consequently, if it later turns out that the arresting officer was reasonably mistaken, it does not make the arrest illegal. Nor does it typically make the officer liable for false arrest. It is just one of those things that can easily happen to anyone.

Examples of information that helps to establish probable cause include:

- A witness is contacted by police who identifies the suspect

- A videotape exists of a crime and is used by police to identify the suspect
- Possession of stolen property or burglary tools on a suspect's person
late at night in a minimally traveled business park where an officer discovers a door ajar
- A dispatcher's broadcast of a physical description of a robbery suspect that closely matches the person who is currently stopped by the police in front of a convenience store i.e. 6'0, 190lbs, long brown hair, small pierced nose ring, orioles baseball hat, and grey deck shoes

**Investigatory Stop (Detention):** An investigatory stop or **temporary detention** of a person by the police is considered by the courts to be a minimal intrusion into a person's routine daily activities and does not constitute an arrest. It is just a brief delay. Most detentions that last approximately **twenty minutes or less** have been deemed to be reasonable by the courts. Some circumstances may warrant a longer detention time, however they must be clearly justified under the circumstances.

A police officer's mere suspicion of a crime, or hunch alone, is not sufficient under the law to support the making of an arrest. However, if while conducting the investigatory stop, an officer's suspicions are eventually confirmed by additional information or evidence, the detention may "ripen" into sufficient probable cause to believe that the person detained has been involved in criminal activity, and the officer is justified at that time in making an arrest of the detained person.

What are some of the circumstances today under which the courts have indicated that an investigatory stop is warranted or acceptable?

- If your car had tinted windows that are too dark such that the police cannot see who is riding inside of the vehicle, they could stop your car to see who is inside of the car. However, the Maryland Courts have recently ruled this practice to be illegal.
- If you have a container such as a backpack in your car during a traffic stop, the police upon making certain observations or movements of the driver or passenger can now search the backpack for evidence
- To verify the Vehicle Identification Number (VIN) of a vehicle if the VIN is not readily identifiable, or appears to have been altered

In recent years, the courts have expanded the authority of police officers to make the driver and all passengers exit from a vehicle during a traffic stop. However, a recent Maryland court decision may alter this practice as to the generalized search of innocent passengers who are not suspected of involvement in any criminal activity.

**Police Contact(s)**: Police officers do not have to have any particular reason or legal basis for simply contacting and talking to a person with whom they want to speak in public. Most people generally want to comply with an officer's request to talk with them despite having a slight feeling of discomfort that they may not be free to leave. Generally these contacts are useful for police in gathering information, developing confidential informant sources, and establishing a familiar presence with people in the community. Some police departments have utilized a "knock and talk" approach to contact people for the purposes of investigating possible criminal activity. During the knock and talk there is an expectation that contacted persons will volunteer useful information that will lead to arrests, intelligence gathering or serve as a deterrent to future criminal conduct by known offenders in the community.

## What Happens Within The First 24 Hours of Being Arrested in Maryland?

Within **24 hours** of an arrest, law enforcement officers in Maryland are required to take an arrestee before a **District Court Commissioner**. If a person is arrested in
on the weekend in Baltimore City, it may take them until Monday to see a District
Court Commissioner at a central lockup facility.

The **District Court Commissioner** will:

- Determine if **probable cause exists** to issue a **Statement of Charges** against the arrestee for the crimes he is alleged to have committed
- **Advise the arrestee** of the charges against him and explain the possible penalties
- Advise the arrestee of his **right to an attorney** who can help him at trial
- Advise the arrestee of his responsibilities to obtain an attorney in a timely manner before trial, or instruct the arrestee to apply to the **Public Defender's Office** for **representation** no later than 10 days before trial
- Decide whether the arrestee should be released his **own recognizance** [signature promise to appear in court] pending trial and based upon the severity of the case facts
- **Determine** what the arrestee's **bail** should be, or set a "no bail" status to prevent the arrestee's release

During the **intake interview** with the District Court Commissioner, it is important that the arrestee provide the Court Commissioner with enough positive background information from which the arrestee's stability in the following areas can be determined by the District Court Commissioner:

- work history
- time of residence in the community

- strong family ties in the community
- lack of a significant criminal record
- past demonstrated responsibility of showing up for court as required in minor cases such as traffic offenses or alcohol citations
- and to establish that the defendant is not a danger to the community

Generally, persons without lengthy records will be released on their own recognizance simply by signing a written promise to appear in court as notified, without necessity of posting bail.

A release on your **own recognizance** is also known as being **OR'd, ROR'd or Recog'd.** The District Court Commissioner, in assessing whether or not the arrestee poses a flight risk, will take particular note of whether the arrestee has any prior failures to appear (FTA's) for minor offenses such as traffic court or alcohol citations.

On busy nights when there is a lot of police activity in the community, or when the police are short staffed, it may take the arrestee longer than 24 hours before he is taken before a District Court Commissioner. It is also interesting to note that District Court Commissioners are not judges.

## Traffic Stops

Traffic stops represent the most likely contact event that the average resident will have with a police officer in life. Consequently, many people will form their opinions about the police based in part upon the number of times they have been stopped, the reasons for each traffic stop, the respectful manner in which the officer treats them, and whether or not they receive a warning or traffic citation.

Therefore, it is important for us to review the many options that police officers in Maryland have to use their discretion to issue a citation or to make and arrest in a traffic matter.

## Police Arrest Powers-Traffic Stops

### Maryland Traffic Arrest Statute

#### § 26-202. POWER OF ARREST- MD. TRANSPORTATION CODE ANN.
(2007)

(a) In general. -- A police officer **may arrest** without a warrant a person for a violation of the Maryland Vehicle Law, including any rule or regulation adopted under it, or for a violation of any traffic law or ordinance of any local authority of this State, if:

(1) The person has committed or is committing the violation **within the view or presence of the officer,** and the violation is any of the following:

(i) A violation of § 21-1411 or § 22-409 of this article, relating to vehicles transporting hazardous materials; or

(ii) A violation of § 24-111 or § 24-111.1 of this article, relating to the failure or refusal to submit a vehicle to a weighing or to remove excess weight from it;

(2) The person has committed or is committing the violation within the view or presence of the officer, and either:

(i) The person **does not furnish satisfactory evidence of identity**; or

(ii) The officer has reasonable grounds to believe that the person will disregard a traffic citation;

(3) The officer has probable cause to believe that the person has committed the violation, and the violation is any of the following offenses:

(i) Driving or attempting to **drive while under the influence of alcohol**, while impaired by alcohol, or in violation of an alcohol restriction;

(ii) Driving or attempting to drive while impaired by any drug, any combination of drugs, or any combination of one or more drugs and alcohol or while impaired by any controlled dangerous substance;

(iii) Failure to stop, give information, or render reasonable assistance, as required by §§ 20-102 and 20-104 of this article, in the event of an accident resulting in bodily injury to or death of any person;

(iv) **Driving or attempting to drive a motor vehicle while the driver's license or privilege to drive is suspended or revoked;**

(v) Failure to stop or give information, as required by §§ 20-103 through 20-105 of this article, in the event of an accident resulting in damage to a vehicle or other property;

(vi) Any offense that caused or contributed to an accident resulting in **bodily injury to or death of any person;**

(vii) Fleeing or **attempting to elude a police officer**; or

(viii) Driving or attempting to drive a vehicle in violation of § 16-101 of this article;

(4) The person is a **nonresident** and the officer has probable cause to believe that:

(i) The person has committed the violation; and

(ii) The violation contributed to an accident; or

(5) The officer has probable cause to believe that the person has committed the violation, and, subject to the procedures set forth in § 26-203 of this subtitle, the person is issued a traffic citation and refuses to acknowledge its receipt by signature.

(b) Manner of arrest. -- **An arrest under this section shall be made in the same manner as, and without more force than, in misdemeanor cases.**

(c) Arrested person to be taken before commissioner; exception. -- A person arrested under this section shall be taken without unnecessary delay before a District Court commissioner, as specified in § 26-401 of this title, unless the arresting officer in his discretion releases the individual upon the individual's written promise to appear for trial.

## Maryland State Police Officer Qualifications

Persons desiring to become Maryland State Police Officers must meet the following minimum qualifications:

- Be a United States Citizen
- Possess a High School Diploma or GED
- Possess a valid Driver's License in any state and a satisfactory driving record
- Binocular far and near visual acuity, with or without correction, must be 20/20
- Be at least 20 years of age but not older than 59
- Be in excellent physical condition
- Have a good reputation and sound moral character
- Not have any current court orders relating to domestic violence
- Be truthful in every stage of the application process

Most Maryland police departments have similar job qualifications for police officer recruits, with a few jurisdictions requiring 60 college credits such as Montgomery County and Howard County, Maryland.

The future job market for police officers is very bright because of the large number of retiring ranking officers in many police departments nationwide. Consequently, the fierce competition between police agencies seeking qualified police recruits has already begun. As a result, this is a great time for students to start an exciting career in law enforcement to make a valuable contribution to the safety of their communities.

# CHAPTER FOUR

# RULE OF LAW

## Crime

**Crime Defined:** A crime can be generally defined as **an act or omission** in violation of the law for which a punishment is prescribed. A crime can also result from a **failure to act when there is a duty to act**, typically due to the existence of some special relationship i.e. parent-child, or a due to a special contractual or fiduciary (trust) relationship i.e. caretaker-patient, babysitter, etc. A crime can also be described in the following ways:

- A crime is a **"public offense"** for which the law prescribes a punishment or sanction. A public offense involves the violation of a law enacted by a governmental authority [i.e. state, federal, local or administrative law making bodies]

- A crime is an **offense** against **society**.

## *"A Rule of Thumb Formula"*: To Determine When a Crime Exists

A simple way to determine when a crime exists is to use this basic formula: **ACT + INTENT = CRIME**

Generally to have a crime, the elements of the illegal conduct (guilty act or actus reus) and the criminal intent (guilty mind or mens rea) must **coexist at the same time** in order to establish criminal responsibility for violating the Maryland criminal law, particularly in specific intent crimes.

If **one** of the **elements** is **missing**, you usually won't have a crime. However, even if you don't have a crime, the same conduct may nevertheless constitute some other recognized form of prohibited conduct that results in a **"civil wrong"** or private harm known as a tort.

In **tort law**, a violation occurs for a **civil wrong** that is recognized by Maryland law and the remedy is normally to file a lawsuit for **monetary damages** or other injunctive relief for the private wrong or harm.

However, in some **general intent crimes**, the **rule of thumb formula** may not apply because the only intent that must be established by the Prosecutor to obtain a conviction in a general intent crime case is that the defendant had the **intent to do the illegal or prohibited conduct**. In other words, the Prosecutor would only have to prove that h**e meant to do what he did**.

An example: One example would be in the case of a **simple battery** which can result from the harmful, unconsented to, or offensive touching of another. In this case the offending person's intent may be immaterial. The fact that they touched someone in an offensive matter is sufficient..

## Substantive Criminal Law:

**Substantive criminal law** can be viewed as society's orders and as it's determination of what is considered to be **undesirable, bad, prohibited or illegal conduct** in society. Usually substantive law is represented by the state, local or federal criminal statutes or ordinances that are enacted by a legislative body, are punishable by a legally prescribed sanction, and are found in many penal codes.

The Maryland substantive criminal law is principally found in the **Maryland Criminal Law Annotated Code of Maryland**. There is no Maryland Penal Code volume. The Maryland Criminal Law Annotated Code prescribes prohibited conduct and the associated maximum fines, penalties and imprisonment for specific criminal offenses. All students studying criminal justice should become familiar with
the Maryland Criminal Law Annotated Code that can be found both in the library and online.

## Procedural Criminal Law:

The **Procedure criminal law** can be viewed as the **"rules of the game"** that tell the various participants in the criminal justice system **(i.e. police, courts, corrections)** how to treat the people who are accused of violating society's established laws. It also let's the accused defendant know how he can expected to be treated by the criminal justice system if he is accused of violating the law.

The principal focus of **procedural criminal law** is **fairness** for the **accused**, and provides the following protections:
- the receipt by the accused of a **notice of the charges** filed against him
- allows for the defendant to have an opportunity to **appear, be heard and to defend** against the charges
- **before** he is **deprived** of a substantial constitutional rights related to his
  *freedom, property, or a constitutionally protected interest*

The Maryland procedural criminal law is principally found in the Maryland Criminal Procedure Annotated Code Article, in the Maryland Courts and Judicial Proceedings Annotated Code Article, and in the Maryland Rules Annotated Code. Each of these sources of procedural criminal law must be read together and in conjunction with the Maryland state constitution and the

U.S. Constitution to ascertain the true meaning and intent of Maryland procedural criminal law.

# CHAPTER FIVE

# COURTS & COURT PROCEDURES

## INSIDE THE COURTROOM

**Prosecutor** – The **State's Attorney** for each county or **Baltimore City** is the primary state representative responsible for prosecuting criminal offenses on behalf of the State of Maryland when a violation of the law has occurred. The **Prosecutor** has the sole authority and discretion to determine if charges will be filed, what charges will be filed, when the charges will be filed, and what if any criminal charges against a defendant will be dismissed. A judge has no authority to force a Prosecutor to either bring charges or to drop charges against an individual. Consequently, Prosecutors have a lot of power to influence how the criminal justice process works.

**Nolle Prosequi (Nol Pros)**

When criminal charges are **dropped** or **dismissed** against an individual, a **Nolle Prosequi (Nol Pros)** is entered on the court record and the proceeding is concluded. No further activity will occur in the case. In some jurisdictions closer to the Washington,DC - Maryland boundary such as Prince George's County and Montgomery County, a dismissal of criminal charges by the Prosecutor is routinely referred to as a **"Nollie"**.

According to the official **Maryland District Court** criminal case statistics for **January 2007** approximately **7,415** cases were entered as **"Nolle Prosequi"** out of a total of **15, 807** criminal cases that were filed statewide. This data represents a *statewide dismissal rate* of approximately **46% percent** of the criminal cases filed in **a single month**.

What, if anything, does this one month dismissal figure say about the Maryland criminal justice system?

The Prosecutor also has the sole responsibility to bring cases to trial in a timely manner that conforms to the Maryland and federal constitutional speedy trial rights of the defendant. Prosecutors also exist in other Maryland state agencies including the **Attorney General's Office** and the **State Prosecutor's Office**.

The **Attorney General's Office** has generally focused on **white collar crimes** in the past, while the **State Prosecutor's Office** has focused on crimes related to **government officials**, and persons holding political office. Nevertheless, over time, both agencies have expanded the scope of the matters that they handle for prosecution. In other states prosecutors are known as District Attorneys (DA's), Commonwealth's Attorney (Virginia), and Corporation Counsel (Washington, D.C.).

**Stet (Stet Docket)**

In Maryland a case can also be placed on the **"Stet" Docket** ("stetted") which means that the case is removed from the active court docket of cases that are scheduled for trial and placed on an **("inactive")** court docket of cases that have no scheduled trial date. These inactive cases are essentially held in suspense by the court and are *similar to placing the cases on a shelf.* **A stet is not a conviction, is not a dismissal, and is not an acquittal.**

Consequently, a **stet** is still considered to be an **"open charge"** and oftentimes must be completely resolved by dismissal or trial for certain individuals who are being considered for admission into the military, or are applying for jobs that require a criminal licensing background check such as police officers, attorneys, certified public accountants, etc..

A Prosecutor can only enter a **stet** on the court record with the **consent of the defendant,** because in order for the defendant to receive a stet, **he must give up his right to demand a speedy trial**. Eventually, if the defendant does not get into future legal trouble, the "stetted" case will have one of the following outcomes:

- **In Year 1:** Either the Defendant or the Prosecutor can ask for the case to be **reopened for any reason**, and the case will be reset for trial.

- **In Year 2:** The case cannot be reopened by the Prosecutor unless **"good cause" can be shown** to the court such as the Defendant's subsequent involvement in a similar criminal conduct or cases (i.e. drug possession or drug distribution).

- **In Year 3:** The case cannot be reopened by the Prosecutor, and after the expiration of **three years** the case can be **expunged,** meaning that the defendant's criminal record can be **sealed** so that no one has access to the case charging documents, evidence, or hearing transcript.

**Expungement of Criminal Records**

In order for the Defendant to get his record expunged in Maryland, he must give up his rights to civilly sue the police officer, police department, judge, and prosecutor who have been involved with the prosecution of his criminal case. This is a unique process in Maryland that forces a person who may be wrongfully arrested and charged with a crime to choose between having an **open public arrest record** which may erroneously indicate his involvement in criminal activity **versus** having the erroneous **criminal record sealed** by the Court by giving up his right to pursue damages against the government for wrongful arrest and prosecution. However, the Maryland expungement process is very beneficial in that gives someone who has committed a one-time offense, or a minor crime, an opportunity to make a new start unhindered by a criminal conviction while also striking a balance between promoting the continuous operation of the criminal justice system unhindered by friviolus civil lawsuits.

**Criminal Procedure - Expungement of Police Records - Arrest Without Charge**

On October 1, 2007 a new Maryland law House Bill 10 (HB 10) went into effect changing the Criminal Procedure Article to require the **automatic expungement** of criminal records when no criminal charges are filed against the arrestee.

This change in the law was prompted largely because of public outcry over massive *quality of life –type offense* arrests in Baltimore City such as loitering, disorderly conduct, and other minor offenses that often resulted in dropped criminal charges for a large number percentage of the arrestees.

The law repeals certain existing provisions related to the initiation of an expungement request of a specified police record that originally required a prior written notice or otherwise prohibited the filing of the expungement request before the passage of a certain period of time (i.e. typically 3 years). These changes are expected to make it better for those persons who are arrested, not charged, and are seeking employment that requires the total lack of any criminal record.

**Defense Attorney** – The **Defense Attorney** is the lawyer responsible for representing a criminal defendant at trial. The Defense Attorney's **primary allegiance** and responsibility is to her **client.** She must vigorously and zealously represent her client within the ethical boundaries of the law. Generally, the defense counsel's goal at trial is to make the Prosecutor prove each and every element of the offenses with which the defendant is charged beyond a reasonable doubt. Of course, no such level of proof is required if the defendant plea bargains his case away prior to trial. The Defense Attorney can be **private counsel, a panel attorney** who is a member of the private bar but offers his services to indigent defendant's at a reduced fee, or can be **Public Defenders**.

**6th Amendment Right to Counsel** – The **6th Amendment** to the U.S. Constitution guarantees a criminal defendant the right to the effective assistance of competent legal counsel at trial at public expense. It **does not**

**guarantee** the defendant the right to the effective assistance of the <u>best</u> **competent legal counsel** available in the local jurisdiction, which in some instances may certainly be an attorney in private practice but may also be an attorney working in a public agency such as the **Maryland Public Defender's Office.**

Many private defense attorneys in Maryland have either worked as Public Defenders or Assistant State's Attorneys earlier in their legal careers. Others have worked as law clerks, Assistant Attorneys General or entered the private practice of law directly out of law school.

**Public Defenders** – Public Defenders from the Maryland Public Defender's Office represent **"indigent"** defendants who are charged with criminal offenses in the State of Maryland. Indigent generally means that the defendant has limited income, no income, or limited financial resources from which to hire a private attorney for his defense, and qualifies for representation by a Maryland Assistant Public Defender at trial.

**Maryland Public Defenders** are lawyers who have chosen public service as their professional legal career. Most are caring, dedicated and hard working individuals who work under sometimes *crushing caseloads* with responsibility for handing many complex and **socially unpopular criminal offenses** such as murder, child abuse, rape, robbery, drugs, domestic violence, assault, battery, and theft. In **Maryland, approximately 90%** percent of all criminal defendants are represented by the Office of the **Public Defender**. In essence, the **cases** handled by the Maryland Public Defender's Office **play a significant role in shaping the criminal case law in Maryland.**

**Judge** – At trial, a Judge has the responsibility of ensuring that the defendant receives a fair trial, to rule on the admissibility of evidence during trial, to instruct the jury, and to manage the conduct of the lawyers at trial. A judge must be an **impartial arbiter** and must not advocate for either the prosecution or defense at trial.

Initially, most **Judges** in Maryland are **appointed** by the Governor. Appointed Circuit Court Judges must run for election during the next statewide election cycle following their initial appointment in order to receive a **15 year** service term on the bench. A judge's personal self-control, respect for the litigants, balanced demeanor and reputation for fairness are keen qualities that the public, litigants, and lawyers expect and appreciate in the courts.

**District Court Judges** serve for **10 years** terms after their initial appointment by the Governor. They must run for re-election after the 10 year term concludes. Generally, Maryland **lawyers** who desire to become judges must practice law for at least
**5 years** before their appointment to the bench. It would seem logical that good and respected local lawyers would probably make good judges, and should thereby enjoy some advantage during the judicial selection process. However, the competitive jockeying that takes place between lawyers

seeking the limited number of available local judgeships makes it more likely that only judicial applicants who can figure out a way to effectively **get their name before the Governor** are more likely to be appointed as new judicial openings occur.

## THE CRIMINAL TRIAL

**Standard of Proof** – In a criminal trial, the Prosecutor must prove his case *beyond a reasonable doubt.* This means that the **Prosecutor** must convince a jury, or a judge during a bench trial, **to a degree of certainty that leaves no reasonable doubt** in the jury or judge's mind, that the defendant committed the crime. A **reasonable doubt** is a doubt based upon some reason that is generated by a careful review and consideration of the facts of the case presented [only] at trial. Although it is a very high standard of proof for the prosecution, the Prosecutor does not have to prove his case beyond all doubt, or to any mathematical degree. Conversely, the defendant does not have any burden to prove his innocence at trial (with some exceptions such as in a receipt of stolen property case) because a criminal defendant is presumed to be innocent until proven guilty by the Prosecutor at trial. If the Prosecutor does not prove his case at trial, the defendant must be found not guilty under our due process system of justice.

### The Maryland Criminal Trial Sequence

A Maryland criminal trial will typically occur in the following sequence:

**Pretrial Motions** – Motions are requests for the court to issue a written order for specific relief being requested by the Prosecution or Defense. They may also be requests for the discovery of trial evidence, or for information in the possession of the Prosecutor or the Defense Attorney prior to trial.

### Types of Motions

**Motion for Discovery**- A **Discovery Motion** requests disclosure of all of the information that the **Prosecutor** intends to use at trial and may consist of one or more of the following:
- Production of Evidence
- Production of Chemical Tests
- Chain of Custody
- Defendant's Statements/Confessions
- Photos
- Fingerprints
- DNA test results
- Electronic Evidence (Computer Forensics)
- Witness Statements
- Police Reports
- Expert Witness Reports

- Police Lineup Identifications

**Motion to Suppress** – A **Suppression Motion** requests the court to exclude any illegally seized evidence from admission into evidence at trial. Illegally seized evidence is normally considered any evidence that is obtained in violation of the Defendant's rights under the $4^{th}$, $5^{th}$, $6^{th}$ Amendments to the U.S. Constitution, a State Constitution or a statute such as a wiretapping law that causes the evidence to be excluded from the defendant's trial.

**Motion in Limine** – A **Motion in Limine** is a very powerful tool that is used by Prosecutor or Defense Attorney to get the court to exclude testimony or evidence that may prove damaging to either sides' case because it either lacks relevancy, is not authenticated (verifiable) by a witness with actual personal knowledge about the items, events or circumstances about which he is being called to testify, or because of some other disqualifying rules or court decisions that prohibit the introduction of the evidence at trial. It is essentially a motion by one or both of the attorneys to *"keep the bad stuff out"* of court because the evidence is not probative of a contested issue in the case and it is likely that the evidence will hurt their case in front of a jury or trier of fact if admitted.

**Opening Statement** – At trial both the **Prosecutor and Defense Attorney** have an opportunity to make **an opening statement** to the jury for the purposes of explaining their version or theory of what the evidence will be expected to show. The opening statements of the attorneys **are not evidence** in the trial. They are merely preliminary statements or theories about the case that prepare the jury for what they can expect to hear and observe during the trial.

**Direct Examination** – A **Direct Examination** occurs where each side puts **their own witness** on the stand and asks questions of their own witness without leading them to the answer by asking questions that either suggest the answer, or simply require the witness to answer using a "yes" or "no" response. Generally, a witness on direct examination just tells their story about the facts based upon personal knowledge.

**Cross Examination** - **Cross Examination** occurs where each side asks questions of **the other side's witness**. The purpose of cross examination is to **aid the jury or judge in getting to the truth**. Each side **can ask leading questions** of the witness as a **matter of right**. Leading questions:

- allow the jury to observe whether the witness knows what he says he knows
- allow the jury to observe whether the witness saw what he said that he saw
- allow the jury to observe the witnesses' demeanor on the stand

- help to cast doubt on a witness' credibility and believability such that a jury should not trust his testimony or give any weight to what the witness says

**Motion for Judgment of Acquittal** (MJOA) – The **Defense Attorney** is required to make a **Motion for Judgment of Acquittal (MJOA)** to request the charges to be dismissed against his client at **two separate times** during the trial:

- after **all** of the **Prosecutor's evidence** has been presented at trial, and again;
- after **all** of the **evidence** in both the **Prosecutor and Defense** Attorney's case has been presented at trial

Since there is usually some evidence presented by the Prosecutor at trial in each case from which the jury may conclude that the Defendant was somehow **involved** in criminal activity, the **courts normally deny** the Defense Motion for Judgment of Acquittal in approximately **98% percent** of all criminal cases.

**Jury Instructions** – The Judge instructs the jury on law that is applicable to the case after all of the evidence has been presented from the Prosecutor and Defense Attorney.

**Closing Argument** – Both the **Prosecutor** and **Defense Attorney** make closing arguments to the jury to **discuss what they feel the evidence has shown** during the trial. Each side requests the jury to rule in their favor. For example: The Prosecutor argues for a guilty finding [conviction] and the Defense Attorney argues for a not guilty finding [acquittal]. Since the **Prosecutor** has the **burden** of proving the Defendant's guilt beyond a reasonable doubt, he gets to **argue** to the jury **first and last**. The **Defense Attorney** only gets to argue to the jury **once**.

**Prosecution Rebuttal Argument** – The Prosecutor gets to make the last argument to the jury before the jury retires to deliberate.

**Verdict** – The Jury must return a **unanimous** verdict of guilty or not guilty in Maryland. If they can agree of a verdict, then a "hung" jury occurs. If the jury hangs, the defendant can be retried again and again by the Prosecutor as many times as desired, until the jury either returns a conviction or an acquittal
.

**Sentencing** - If the defendant is found guilty at trial, the court may order a **Presentence Investigation Report (PSI)** to learn more about the defendant's personal, emotional, family, educational, and occupational background in addition to the defendant's prior criminal history. The PSI will provide a range of recommended sentences for the court to use as guidelines based upon a **numerical scoring matrix** that corresponds to a **range of pre-**

**established punishments** from probation to jail time to be served by the defendant.

In **Maryland**, the sentencing guidelines are **discretionary and advisory only**, not mandatory pursuant to **Maryland Criminal Procedure Code Ann. § 6-216.** This means that a judge does not have to follow the sentencing guidelines and can go below or above the recommended guideline sentence in a criminal case. However, the judge is typically required to explain why he or she departed from the recommended guidelines. Chapter 6 of this supplement will explore sentencing guideline considerations in greater detail.

# CHAPTER SIX

# MARYLAND CORRECTIONS

**Maryland Probation & Parole Prison Characteristics Data**

**Maryland Probation/Parole Facts at a Glance:**

As of December 2006, the Maryland Division of Parole and Probation had **67,519** Probationers and Parolees under supervision in the following categories:

Probation Criminal Supervision - **41,791** (62%)

Probation Drinking Drivers - **15,786** (23%)

Mandatory Supervision - **5,163** (8%)

Parole - **4,579** (7%)

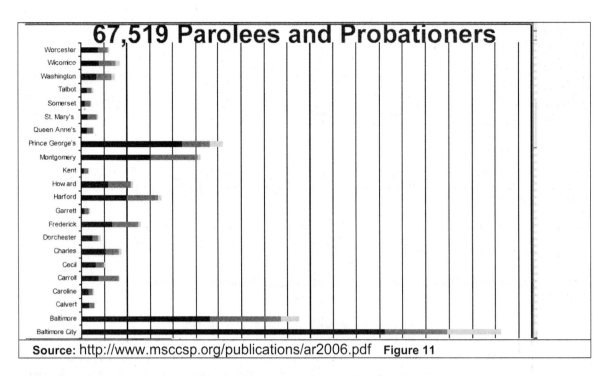

## 67,519 Parolees and Probationers

Worcester
Wicomico
Washington
Talbot
Somerset
St. Mary's
Queen Anne's
Prince George's
Montgomery
Kent
Howard
Harford
Garrett
Frederick
Dorchester
Charles
Cecil
Carroll
Caroline
Calvert
Baltimore
Baltimore City

Source: http://www.msccsp.org/publications/ar2006.pdf    Figure 11

## Maryland Prison Populations Data

| Characteristics of Supervised Populations – Maryland Prisons | | | |
|---|---|---|---|
| | Institutional Populations | | |
| | DPDS | Division of Corrections (DOC) | Patuxent Institute |
| **Gender** | | | |
| Male | **85.9%** | **95.4%** | **84.2%** |
| Female | 14.1% | 4.6% | 15.8% |
| **RACE** | | | |
| Black | **87.4%** | **75.8%** | **78.5%** |
| White | 11.5% | 23.8% | 21.5% |
| Other | 1.1% | 0.4% | 0.0% |
| **Offense Type** | | | |
| Violent | 19.1% | 45.3% | **83.2%** |
| Non-violent | **80.9%** | **54.7%** | 16.8% |
| **Total Population** | 3,385 | 22,972 | 391 |

| | Community Supervision Populations - Maryland | | | |
|---|---|---|---|---|
| | Probation | Parole | Mandatory | DDMP |
| **Gender** | | | | |
| Male | **81.9%** | **88.9%** | **95.3%** | **81.1%** |
| Female | 18.1% | 11.1% | 4.7% | 18.9% |
| **Race** | | | | |
| Black | 58.5% | 75.3% | 73.6% | 17.3% |

| | | | | |
|---|---|---|---|---|
| White | **41.1%** | 24.5% | 26.2% | **81.4%** |
| Other | 0.4% | 0.2% | 0.2% | 1.3% |
| **Offense Type** | | | | |
| Violent | 8.2% | 20.7% | 34.6% | 0.0% |
| Non-violent | **91.8%** | 79.3% | 65.4% | **100.0%** |
| **Total Population** | 77,644 | 6,503 | 7,928 | |
| | | | | **20,175** |

Notes: Institutional populations are based on counts as of June 30, 2005. 155 sentenced offenders held at DPDS are included in the DOC count. Patuxent Institution data does not include 39 re-entry facility offenders.

Community supervision populations are based on counts of cases being administered by the DPP as of June 30, 2005. The DPDS Pretrial Release Services Program supervised 1,359 defendants as of June 30, 2005 for which descriptive data is unavailable.

All percentages were obtained from automated information systems and may differ from other published reports.

Source: http://www.msccsp.org/publications/ar2006.pdf

**Figure 12**

## Maryland Sentencing Guidelines Worksheet

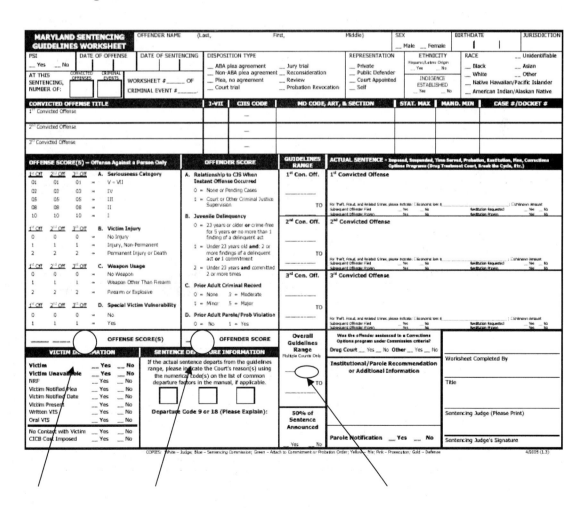

57

Offense Score   &   Offender Score       **Figure 13**       Overall Guidelines
Are Used With Scoring Matrix                            Sentence Range

**Source: http://www.msccsp.org/guidelines/matrices.html#person**

Both Scores Are Entered on Sentencing Guidelines Worksheet

Recommended Sentence for Offense Score "2 "& Offender Score "1" = P-1Yr

GUIDELINES SCORING MATRICES
# Person Offenses

| | 0 | 1 | 2 | 3 | 4 | 5 | 6 | 7+ |
|---|---|---|---|---|---|---|---|---|
| 1 | P | P | P-3M | 3M-1Y | 3M-18M | 3M-2Y | 6M-2Y | 1Y-3Y |
| 2 | P-6M | P-1Y | P-18M | 3M-2Y | 6M-3Y | 1Y-5Y | 18M-5Y | 3Y-8Y |
| 3 | P-2Y | P-2Y | 6M-3Y | 1Y-5Y | 2Y-5Y | 3Y-7Y | 4Y-8Y | 5Y-10Y |
| 4 | P-3Y | 6M-4Y | 1Y-5Y | 2Y-5Y | 3Y-7Y | 4Y-8Y | 5Y-10Y | 5Y-12Y |
| 5 | 3M-4Y | 6M-5Y | 1Y-6Y | 2Y-7Y | 3Y-8Y | 4Y-10Y | 6Y-12Y | 8Y-15Y |
| 6 | 1Y-6Y | 2Y-7Y | 3Y-8Y | 4Y-9Y | 5Y-10Y | 7Y-12Y | 8Y-13Y | 10Y-20Y |
| 7 | 3Y-8Y | 4Y-9Y | 5Y-10Y | 6Y-12Y | 7Y-13Y | 9Y-14Y | 10Y-15Y | 12Y-20Y |
| 8 | 4Y-9Y | 5Y-10Y | 5Y-12Y | 7Y-13Y | 8Y-15Y | 10Y-18Y | 12Y-20Y | 15Y-25Y |
| 9 | 5Y-10Y | 7Y-13Y | 8Y-15Y | 10Y-15Y | 12Y-18Y | 15-25Y | 18Y-30Y | 20Y-30Y |
| 10 | 10Y-18Y | 10Y-21Y | 12Y-25Y | 15Y-25Y | 15Y-30Y | 18Y-30Y | 20Y-35Y | 20Y-L |
| 11 | 12Y-20Y | 15Y-25Y | 18Y-25Y | 20Y-30Y | 20Y-30Y | 25Y-35Y | 25Y-40Y | 25Y-L |
| 12 | 15Y-25Y | 18Y-25Y | 18Y-30Y | 20Y-35Y | 20Y-35Y | 25Y-40Y | 25Y-L | 25Y-L |
| 13 | 20Y-30Y | 25Y-35Y | 25Y-40Y | 25Y-L | 25Y-L | 30Y-L | L | L |
| 14 | 20Y-L | 25Y-L | 28Y-L | 30Y-L | L | L | L | L |
| 15 | 25Y-L | 30Y-L | 35Y-L | L | L | L | L | L |

**P=Probation, M=Months, Y=Year, L=Life**
**Figure 14**

## Maryland Sentencing Guideline Scoring Matrix — Drug Offenses

## Drug Offenses

| | 0 | 1 | 2 | 3 | 4 | 5 | 6 | 7 |
|---|---|---|---|---|---|---|---|---|
| VII | P | P | P | P-1M | P-3M | P-6M | 3M-6M | 6M-2Y |
| VI | Available for future use. There are currently no seriousness category VI drug offenses. | | | | | | | |
| V | P-6M | P-12M | 3M-12M | 6M-18M | 1Y-2Y | 1.5Y-2.5Y | 2Y-3Y | 3Y-4Y |
| IV | P-12M | P-18M | 6M-18M | 1Y-2Y | 1.5Y-2.5Y | 2Y-3Y | 3Y-4Y | 3.5Y-10Y |
| III-A (Marijuana import 45 kilograms or more, and MDMA over 750 grams) | P-18M | P-2Y | 6M-2Y | 1Y-4Y | 2Y-6Y | 3Y-8Y | 4Y-12Y | 10Y-20Y |
| III-B (Non-marijuana and non-MDMA, Except Import) | 6M-3Y | 1Y-3Y | 18M-4Y | 3Y-7Y | 4Y-8Y | 5Y-10Y | 7Y-14Y | 12Y-20Y |
| III-C (Non-marijuana and non-MDMA, Import) | 1Y-4Y | 2Y-5Y | 3Y-6Y | 4Y-7Y | 5Y-8Y | 6Y-10Y | 8Y-15Y | 15Y-25Y |
| II | 20Y-24Y | 22Y-26Y | 24Y-28Y | 26Y-30Y | 28Y-32Y | 30Y-36Y | 32Y-37Y | 35Y-40Y |

**Figure 15**

**Source**: Figure 14 and Figure 15-http://www.msccsp.org/guidelines/matrices.html#person

# Maryland Sentencing Guideline Scoring Matrix — Property Offenses

## Property Offenses

|  | 0 | 1 | 2 | 3 | 4 | 5 | 6 | 7+ |
|---|---|---|---|---|---|---|---|---|
| VII | P-1M | P-3M | 3M-9M | 6M-1Y | 9M-18M | 1Y-2Y | 1Y-3Y | 3Y-5Y |
| VI | P-3M | P-6M | 3M-1Y | 6M-2Y | 1Y-3Y | 2Y-5Y | 3Y-6Y | 5Y-10Y |
| V | P-6M | P-1Y | 3M-2Y | 1Y-3Y | 18M-5Y | 3Y-7Y | 4Y-8Y | 8Y-15Y |
| IV | P-1Y | 3M-2Y | 6M-3Y | 1Y-4Y | 18M-7Y | 3Y-8Y | 5Y-12Y | 10Y-20Y |
| III | P-2Y | 6M-3Y | 9M-5Y | 1Y-5Y | 2Y-8Y | 3Y-10Y | 7Y-15Y | 15Y-30Y |
| II | 2Y-5Y | 3Y-7Y | 5Y-8Y | 5Y-10Y | 8Y-15Y | 10Y-18Y | 12Y-20Y | 15Y-40Y |

**Source:** Figure 14 and Figure 15-http://www.msccsp.org/guidelines/matrices.html#person

**Figure 16**

# Maryland Sentencing Seriousness Classifications

## Maryland Guidelines Seriousness Category Classifications

The following Offense Seriousness Category Reclassifications were approved by the Maryland State Commission on Criminal Sentencing Policy for adoption in February 2007 into the Code of Maryland Administrative Regulations (COMAR) to bring them in line with other similar offenses.

majority vote, the Commission approved the revised seriousness categories noted in Table 4.

These changes were submitted to the COMAR and their adoption is expected in February 2007.

**Table 4. Offense Seriousness Categories Modified by the SCCSP in 2006, Pending Adoption in COMAR.**

| Statute | Offense | Prior Seriousness Category | New Seriousness Category |
|---|---|---|---|
| CR, §3-307(a)(1)[1] | Sex offense, 3$^{rd}$ degree: use of dangerous weapon; suffocate, strangle, disfigure or inflict serious injury; or while aided and abetted by another | V | IV |
| CR, §3-307(a)(2) | Sex offense, 3$^{rd}$ degree: with mentally defective, mentally incapacitated, or physically hapless individual | V | IV |
| CR, §3-321 | Sodomy | V | IV |
| CR, §3-315 | Continuing course of conduct which includes 3 or more acts involving 1$^{st}$ or 2$^{nd}$ degree rape or 1$^{st}$, 2$^{nd}$, or 3$^{rd}$ degree sex offense over a period of 90 days or more with a victim younger than 14 years old | III | II |
| CR, §3-323 | Incest | V | IV |
| CR, §3-324 | Sexual solicitation of a minor | V | IV |
| CR, §3-602 | Child Abuse, sexual | III | II |
| CR, §3-601(b)(2)(i) | Child Abuse, physical, 1$^{st}$ degree | III | II |
| CR, §8-504(b) | Fraudulent statement in application for public assistance | V | IV |

**Source:** http://www.msccsp.org/publications/ar2006.pdf
**Maryland State Commission on Criminal Sentencing Policy 2006 Annual Report**

Figure 17

| COMAR# | Offense Literal | CJIS Code | Source | Felony or Misd. | Max Term | Min Term | Offense Type | Serious. Categ. | Fine |
|---|---|---|---|---|---|---|---|---|---|
| 22 | **Assault and other Bodily Woundings** Assault, 1st degree | 1-1420 | CR, §3-202 | Felony | 25Y ★ | | Person | III | |
| 24 | **Assault and Other Bodily Woundings** Assault, 2nd degree | 1-1415 | CR, §3-203 | Misd. | 10Y | | Person | V | $2,500 |
| 25 | **Assault and Other Bodily Woundings** Female genital mutilation—perform or consent to | 1-0965 1-0970 | HG, §20-603 | Felony | 5Y | | Person | V | $5,000 |
| 25-1 | **Assault and Other Bodily Woundings—Other** Willfully expose to infectious disease | | HG, §18-601 | Misd. | 1Y | | Person | VII | $500 |
| 25-2 | **Assault and Other Bodily Woundings—Other** Knowingly transfer or attempt to transfer HIV virus | 4-7360 | HG, §18-601.1 | Misd. | 3Y | | Person | V | $2,500 |
| 25-4 | **Assault and Other Bodily Woundings—Other** Knowingly and willfully causing another to ingest bodily fluid | 1-0321 | CR, §3-215 | Misd. | 10Y | | Person | V | |
| 25-5 | **Assault and Other Bodily Woundings—Other** Failure to comply with or violation of protective order, 1st offense | 2-0254 | FL, §4-509(a)(1) | Misd. | 90D | | Person | VII | $1,000 |
| 25-6 | **Assault and Other Bodily Woundings—Other** Failure to comply with or violation of protective order, **subsequent** | 2-0254 | FL, §4-509(a)(2) | Misd. | 1Y | | Person | VII | $2,500 |
| 25-7 | **Assault and Other Bodily Woundings—Other** Failure to comply with or violation of peace order | 2-0105 | CJ, §3-1508 | Misd. | 90D | | Person | VII | $1,000 |
| 26 | **Assault and Other Bodily Woundings** Reckless endangerment | 1-1425 1-1430 (from car) | CR, §3-204(a) | Misd. | 5Y | | Person | V | $5,000 |
| 27 | **Assault and Other Bodily Woundings** Cause a life threatening injury by motor vehicle or vessel while under the influence of alcohol | 1-0765 | CR, §3-211(c) | Misd. | 3Y | | Person | VI | $5,000 |
| 28 | **Assault and Other Bodily Woundings** Cause a life threatening injury by motor vehicle or vessel while impaired by alcohol | 1-0770 | CR, §3-211(d) | Misd. | 2Y | | Person | VI | $3,000 |

**Source:** http://www.msccsp.org/publications/ar2006.pdf
**Maryland State Commission on Criminal Sentencing Policy 2006 Annual Report**

**Figure 18**

| COMAR# | Offense Literal | CJIS Code | Source | Felony or Misd. | Max Term | Min Term | Offense Type | Serious. Categ. | Fine |
|---|---|---|---|---|---|---|---|---|---|
| 109 | **Counterfeiting** Falsifying, destroying, concealing, accessing, etc., public records | 2-2504 3-2504 5-2504 | CR, §8-606(b) | Misd. | 3Y | | Property | VII | $1,000 |
| 110 | **Counterfeiting** Forgery—trademark counterfeiting, less than $1,000 | 1-2545 | CR, §8-611(d) | Misd. | 18M | | Property | VII | $1,000 |
| 112 | **Counterfeiting** Possess, utter forged, etc. U.S. currency | 1-0468 | CR, §8-604.1(a) | Misd. | 3Y | | Property | VII | $1,000 |
| 112-1 | **Counterfeiting** Possession of counterfeit items | | CR, §8-601(c)(2) | Misd. | 3Y | | Property | VII | $1,000 |
| 113 | **Counterfeiting** Unlawful possession of forged, etc., motor vehicle title | 1-0468 | CR, §8-603(a) | Misd. | 3Y | | Property | VII | $1,000 |
| 114 | **Counterfeiting** Make, utter, forge, etc. tickets, coupons, tokens, etc. without the authority of the person or corporation issuing, selling, etc.; assist in making, uttering, forging, etc. of tickets, coupons, tokens, etc.; or utter or pass, knowing it to be so made | 3-2502 3-2510 | CR, §8-612(b) | Misd. | 1Y | | Property | VII | |
| 115 | **Credit Card Crimes** Felony Credit Card, greater than $500 | 1-0196 1-0197 1-0198 1-0199 1-0200 3-4125 3-4135 3-4145 3-4155 3-4165 3-4175 | CR, §8-207(b)(1) | Felony | 15Y | | Property | V | $1,000 |
| 116 | **Credit Card Crimes** Credit Cards—possess with unlawful intent a payment device number | 3-2210 | CR, §8-215 CR, §8-216 | Felony | 15Y | | Property | V | $1,000 |
| 117 | **Credit Card Crimes** Misdemeanor Credit Card, less than $500 | 1-2607 1-2399 2-2803 1-2899 1-2605 1-2803 3-4120 3-4130 3-4140 3-4150 3-4160 3-4170 | CR, §8-207(b)(2) | Misd. | 18M | | Property | VII | $500 |

**Source:** http://www.msccsp.org/publications/ar2006.pdf
**Maryland State Commission on Criminal Sentencing Policy 2006 Annual Report**
**Figure 19**

## The Maryland Statute on Parole Eligibility

A review of the following Maryland Parole Eligibility Statute is useful for students studying criminal justice, victims, and defendants in criminal matters because it helps to dispel some of the myths that currently exist about the total amount of time an incarcerated individual will actually serve in Maryland prisons for non-violent and violent crimes. Our hope is that the reader will gain a greater sense of clarity about how sentences are computed to help in better judging some of the more notorious cases today of defendants who are sentenced or released and are routinely featured on state and national television everyday.

§ 7-301. Eligibility for parole - **Md. CORRECTIONAL SERVICES Code Ann. § 7-301 (2007)**

**(a) In general. --**

(1) Except as otherwise provided in this section, the Commission shall request that the Division of Parole and Probation make an investigation for inmates in a local correctional facility and the Division of Correction make an investigation for inmates in a State correctional facility that will enable the Commission to determine the advisability of granting parole to an inmate who:

(i) has been sentenced under the laws of the State to serve a term of 6 months or more in a correctional facility; and

(ii) has served in confinement one-fourth of the inmate's aggregate sentence.

(2) Except as provided in paragraph (3) of this subsection, or as otherwise provided by law or in a predetermined parole release agreement, an inmate is not eligible for parole until the inmate has served in confinement one-fourth of the inmate's aggregate sentence.

(3) An inmate may be released on parole at any time in order to undergo drug or alcohol treatment if the inmate:

(i) is not serving a sentence for a crime of violence, as defined in § 14-101 of the Criminal Law Article;

(ii) is not serving a sentence for a violation of Title 3, Subtitle 6, § 5-608(d), § 5-609(d), § 5-612, § 5-613, § 5-614, § 5-621, § 5-622, or § 5-628 of the Criminal Law Article; and

(iii) has been determined to be amenable to drug or alcohol treatment.

(b) Multiple terms. -- Except as provided in subsection (c) of this section, if an inmate has been sentenced to a term of imprisonment during which the inmate is eligible for parole and a term of imprisonment during which the inmate is not eligible for parole, the inmate is not eligible for parole consideration under subsection (a) of this section until the inmate has served the greater of:

(1) one-fourth of the inmate's aggregate sentence; or

(2) a period equal to the term during which the inmate is not eligible for parole.

**(c) Violent crimes. --**

(1) (i) Except as provided in subparagraph (ii) of this paragraph, an inmate who has been sentenced to the Division of Correction after being convicted of a violent crime committed on or after October 1, 1994, is not eligible for parole until the inmate has served the greater of:

    1. one-half of the inmate's aggregate sentence for violent crimes; or

    2. one-fourth of the inmate's total aggregate sentence.

(ii) An inmate who has been sentenced to the Division of Correction after being convicted of a violent crime committed on or after October 1, 1994, and who has been sentenced to more than one term of imprisonment, including a term during which the inmate is eligible for parole and a term during which the inmate is not eligible for parole, is not eligible for parole until the inmate has served the greater of:

    1. one-half of the inmate's aggregate sentence for violent crimes;

    2. one-fourth of the inmate's total aggregate sentence; or

    3. a period equal to the term during which the inmate is not eligible for parole.

(2) An inmate who is serving a term of imprisonment for a violent crime committed on or after October 1, 1994, shall receive an administrative review of the inmate's progress in the correctional facility after the inmate has served the greater of:

    (i) one-fourth of the inmate's aggregate sentence; or

    (ii) if the inmate is serving a term of imprisonment that includes a mandatory term during which the inmate is not eligible for parole, a period equal to the term during which the inmate is not eligible for parole.

(d) Life imprisonment. --

(1) Except as provided in paragraphs (2) and (3) of this subsection, an inmate who has been sentenced to life imprisonment is not eligible for parole consideration until the inmate has served 15 years or the equivalent of 15 years considering the allowances for diminution of the inmate's term of confinement under § 6-218 of the Criminal Procedure Article and Title 3, Subtitle 7 of this article.

(2) An inmate who has been sentenced to life imprisonment as a result of a proceeding under § 2-303 or § 2-304 of the Criminal Law Article is not eligible for parole consideration until the inmate has served 25 years or the equivalent of 25 years considering the allowances for diminution of the inmate's term of confinement under § 6-218 of the Criminal Procedure Article and Title 3, Subtitle 7 of this article.

(3) (i) If an inmate has been sentenced to imprisonment for life without the possibility of parole under § 2-203 or § 2-304 of the Criminal Law Article, the inmate is not eligible for parole consideration and may not be granted parole at any time during the inmate's sentence.

(ii) This paragraph does not restrict the authority of the Governor to pardon or remit any part of a sentence under § 7-601 of this title.

(4) If eligible for parole under this subsection, an inmate serving a term of life imprisonment may only be paroled with the approval of the Governor.

# CHAPTER SEVEN

# JUVENILE COURTS

## Philosophy of the Juvenile Courts

The **Maryland philosophy of the Juvenile Courts** his similar to courts in other U.S. jurisdictions as to certain guiding principals and non-punitive social purposes as follows:

- that juvenile proceedings are of a special nature designed to meet the problems peculiar to the adolescent

- that the **proceedings** of a juvenile court **are not criminal** in nature and its dispositions are not punishment for crime

- that the juvenile law has as its underlying concept the protection of the juvenile, so that judges, in making dispositions in juvenile cases, think not in terms of guilt, but of the **child's need for protection or rehabilitation**

- that the juvenile act **does not contemplate the punishment** of children where they are found to be delinquent, but rather an attempt to correct and rehabilitate them in "a wholesome family environment whenever possible," although rehabilitation may have to be sought in some instances in an institution

## THE JUVENILE PROCESS:

Juvenile Court proceedings in Maryland are considered to be **civil matters** and are governed by many of the **due process** [fairness] **protections** such as notice, hearings, and standards of proof that are afforded to adults who are faced with charges of violating the criminal law.

### How Juvenile Charges are Filed:
The **Prosecutor** – a State's Attorney in each Maryland county or Baltimore City handles juvenile criminal matters as the State's representative and files a **Petition for a Finding of Delinquency** instead of filing criminal charges against the juvenile, as is done in the adult courts.

### Juvenile Court Judges:
Juvenile Court Judges are known as " **Juvenile Masters**" who preside over the juvenile court case docket and conduct "**adjudication**" hearings.

**Juvenile Masters Court Findings:**
If the Juvenile Court Master determines that the juvenile has committed the violation(s) that have been alleged against her, the Master will make a finding of **"involved"** against the juvenile.

The juvenile court has many disposition options available including:
- Probation
- community service
- mental health counseling
- anger management
- drug treatment to incarceration in a juvenile facility

However, one of the **underlying goals** of the juvenile system is to avoid the stigma of a criminal conviction in order to give the juvenile **an opportunity to grow, and become a productive member of society**

# MARYLAND JUVENILE PROCEEDINGS STATUTE

The following Md. Courts and Judicial Proceedings Articles set forth the manner in which Juvenile Proceedings should be conducted in the Maryland courts.

## Maryland Juvenile Court Judicial Proceedings Statute

### Md. COURTS AND JUDICIAL PROCEEDINGS Code Ann. § 3-8A-02 (2007)

(a) Purposes of subtitle. -- The purposes of this subtitle are:
(1) To ensure that the Juvenile Justice System **balances** the following objectives for children who have committed delinquent acts:

(i) **Public safety** and the protection of the community;

(ii) **Accountability of the child** to the **victim** and the community for offenses committed; and

(iii) Competency and **character development** to assist children in becoming responsible and productive **members of society**;

(2) To hold **parents** of children found to be delinquent **responsible for the child's behavior** and accountable to the victim and the community;

(3) To hold parents of children found to be delinquent or in need of supervision responsible, where possible, for **remedying the circumstances** that required the court's intervention;

(4) To provide for the care, protection, and **wholesome mental and physical development of children** coming within the provisions of this subtitle; and to provide for a program of treatment, training, and rehabilitation consistent with the child's best interests and the protection of the public interest;

(5) To conserve and **strengthen the child's family ties** and to separate a child from his parents only when necessary for his welfare or in the interest of public safety;

(6) **If necessary** to **remove a child from his home**, to secure for him custody, care, and discipline as nearly as possible equivalent to that which should have been given by his parents;

(7) To provide to **children in State care** and custody:

(i) A **safe, humane, and caring environment**; and

(ii) Access to required services; and

(8) To provide judicial procedures for carrying out the provisions of this subtitle.

(b) Construction of subtitle. -- This subtitle shall be liberally construed to effectuate these purposes.

---

## Juvenile Adjudications (Hearings)

Md. COURTS AND JUDICIAL PROCEEDINGS Code Ann. § 3-8A-23 (2007)

(a) In general. --
(1) An **adjudication** of a child pursuant to this subtitle **is not a criminal conviction for any purpose** and does not impose any of the civil disabilities ordinarily imposed by a criminal conviction.

(2) An adjudication and disposition of a child in which the child's driving privileges have been suspended may not affect the child's driving record or result in a point assessment. The State Motor Vehicle Administration may not disclose information concerning or relating to a suspension under this subtitle to any insurance company or person other than the child, the child's parent or guardian, the court, the child's attorney, a State's Attorney, or law enforcement agency.

(3) Subject to paragraph (4) of this subsection, an adjudication of a child as delinquent by reason of the child's violation of the State vehicle laws, including a violation involving an unlawful taking or unauthorized use of a motor vehicle under § 7-105 or § 7-203 of the Criminal Law Article or § 14-102 of the Transportation Article shall be reported by the clerk of the court to the Motor Vehicle Administration, which shall assess points against the child under Title 16, Subtitle 4 of the Transportation Article, in the same manner and to the same effect as if the child had been convicted of the offense.

(4) (i) An adjudication of a child as delinquent by reason of the child's violation of § 21-902 of the Transportation Article or a finding that a child has committed a delinquent act by reason of the child's violation of § 21-902 of the Transportation Article, without an adjudication of the child as delinquent, shall be reported by the clerk of the court to the Motor Vehicle Administration which shall suspend the child's license to drive as provided in § 16-206(b) of the Transportation Article:

1. For 1 year for a first adjudication as delinquent or finding of a delinquent act for a violation of § 21-902 of the Transportation Article; and

2. For 2 years for a second or subsequent adjudication as delinquent or finding of a delinquent act for a violation of § 21-902 of the Transportation Article.

(ii) In the case of a finding, without an adjudication, that a child has violated § 21-902 of the Transportation Article, the Motor Vehicle Administration shall retain the report in accordance with § 16-117(b)(2) of the Transportation Article pertaining to records of licensees who receive a disposition of probation before judgment.

(b) **Adjudication and disposition not admissible as evidence**. -- An adjudication and disposition of a child pursuant to this subtitle are not admissible as evidence against the child:

(1) **In any criminal proceeding prior to conviction**; or

(2) **In any adjudicatory hearing on a petition alleging delinquency**; or

(3) **In any civil proceeding** not conducted under this subtitle.

(c) Evidence given in proceeding under this subtitle inadmissible in criminal proceeding. -- Evidence given in a proceeding under this subtitle is not admissible against the child in any other proceeding in another court, except in a criminal proceeding where the child is charged with perjury and the evidence is relevant to that charge and is otherwise admissible.

(d) **State employment**. -- An adjudication or disposition of a child under this subtitle shall not disqualify the child with respect to employment in the civil service of the State or any subdivision of the State.

---

The following Md. Criminal Procedure Article sets forth the procedures for **transferring cases to juvenile court** where juveniles have been charged as adults in the Circuit Court.

## Juvenile Transfers [From Adult Court]

Md. CRIMINAL PROCEDURE Code Ann. § 4-202 (2007)

### § 4-202. **Transfer of criminal cases to juvenile court**

(a) Definitions. --

(1) In this section the following words have the meanings indicated.

(2) "Victim" has the meaning stated in § 11-104 of this article.

(3) "Victim's representative" has the meaning stated in § 11-104 of this article.

(b) When transfer allowed. -- Except as provided in subsection (c) of this section, a court exercising criminal jurisdiction in a case involving a child may transfer the case to the juvenile court before trial or before a plea is entered under Maryland Rule 4-242 if:

(1) the accused child was at least 14 but not 18 years of age when the alleged crime was committed;

(2) the alleged crime is excluded from the jurisdiction of the juvenile court under § 3-8A-03(d)(1), (4), or (5) of the Courts Article; and

(3) the court determines by a preponderance of the evidence that a transfer of its jurisdiction is in the interest of the child or society.

(c) Transfer prohibited. -- The court <u>may not</u> transfer a case to the juvenile court under subsection (b) of this section if:

(1) the child previously has been transferred to juvenile court and adjudicated delinquent;

(2) the child was convicted in an unrelated case excluded from the jurisdiction of the juvenile court under § 3-8A-03(d)(1) or (4) of the Courts Article; or

(3) the alleged crime is murder in the first degree and the accused child was 16 or 17 years of age when the alleged crime was committed.

(d) Transfer criteria. -- In determining whether to transfer jurisdiction under subsection (b) of this section, the court shall consider:

(1) the age of the child;

(2) the mental and physical condition of the child;

(3) the amenability of the child to treatment in an institution, facility, or program available to delinquent children;

(4) the nature of the alleged crime; and

(5) the public safety.

(e) Study concerning child. -- In making a determination under this section, the court may order that a study be made concerning the child, the family of the child, the environment of the child, and other matters concerning the disposition of the case.

(f) Transfer determination. -- The court shall make a transfer determination within 10 days after the date of a transfer hearing.

(g) Procedures on transfer -- Juvenile court. -- If the court transfers its jurisdiction under this section, the court may order the child held for an adjudicatory hearing under the regular procedure of the juvenile court.

(h) Holding in juvenile facility. --

(1) Pending a determination under this section to transfer its jurisdiction, the court may order a child to be held in a secure juvenile facility.

(2) A hearing on a motion requesting that a child be held in a juvenile facility pending a transfer determination shall be held not later than the next court day, unless extended by the court for good cause shown.

(i) Rights of victims. --

(1) A victim or victim's representative shall be given notice of the transfer hearing as provided under § 11-104 of this article.

(2) (i) A victim or a victim's representative may submit a victim impact statement to the court as provided in § 11-402 of this article.

(ii) This paragraph does not preclude a victim or victim's representative who has not filed a notification request form under § 11-104 of this article from submitting a victim impact statement to the court.

(iii) The court shall consider a victim impact statement in determining whether to transfer jurisdiction under this section.

(j) Disposition by District Court. -- At a bail review or preliminary hearing before the District Court involving a child whose case is eligible for transfer under subsection (b) of this section, the District Court may order that a study be made under the provisions of subsection (e) of this section, or that the child be held in a secure juvenile facility under the provisions of subsection (h) of this section, regardless of whether the District Court has criminal jurisdiction over the case.

# CHAPTER EIGHT

# MARYLAND TERRORISM

SECTION 8: TERRORISM

## Terrorism

**Maryland Terrorism:** Maryland generally follows the **FBI definition of terrorism** which is "the unlawful use of force or violence against persons or property to intimidate, or coerce a government, the civilian population, or any segment thereof,
in furtherance of political or social objectives".

Today, the term "terrorism" encompasses a broad range of potential social behaviors in society, and the courts are constantly grappling with the true meaning of what constitutes terrorist activity in America amidst the few criminal cases that have actually been tried since the 911 attack on the U.S. in 2000.

Although, some Maryland authorities have publicly indicated that the likelihood of a terrorist attack in Maryland as being remote, many police and fire departments have nevertheless incorporated terrorism training into their emergency response preparedness plans.

However, rising population growth, increasing housing density and overburdened highways in many Maryland communities make it a more likely scenario that citizens within each community will be the true first responders called upon to render emergency assistance to their fellow neighbors in the event of an extreme natural disaster or terrorist attack.

Therefore it becomes important and necessary for current criminal justice students to consider what types of daily behavior occurring in the streets of America constitutes "terrorism" or "domestic terrorism" so as to operate with unlawful purposes, against people or property to intimidate.., with political or social objectives?   For example, does gang activity meet the definition of terrorism?

**Source:** Maryland Emergency Management Agency (http://mema.state.md.us)

# CHAPTER NINE

# MARYLAND GANGS

SECTION 9: GANGS

## Youth Gangs

**Youth Gang Defined:** In Maryland, according to the **Maryland Gang Information and Prevention Office**, a youth gang is defined as "a group of three or more
individuals between the ages of **12-24**:
- who engage in delinquent or criminal activity
- to benefit the gang or its reputation
- with an identifying name
- who share common symbols, clothing, graffiti and hand signs
- and regularly meet and/or claim a specific geographic location as their territory

Baltimore, Maryland |

## Professor: Murder trial showed Columbia has a gang problem

Jan 31, 2007 3:00 AM
by Luke Broadwater, The Examiner

**Columbia** - A college professor says a recent murder trial illustrates that Columbia has a noticeable gang problem. ...

"We need to do something about it before [gangs get] out of control," Howard County Community College criminal justice professor Patrick O'Guinn said. ..

..."I don't care if they don't have a formal gang name," O'Guinn said. "If 30 people are looking to fight someone, that's a gang."

Figure 20

Various types of gangs have existed in Maryland for many years and have usually been linked to the drug trade and gun related violence. However, new attention is being focused on Maryland gang activity because nationally recognized gangs originating on the West Coast such as the **Crips, Bloods**, and **MS-13**, are now gaining growing visibility in Maryland communities and prisons in addition to the well known East Coast gangs.

There is no universal definition on what constitutes a **"gang"** and the classification of what constitutes gang activity still varies from state to state. This is so because the labeling of certain activity as gang-related provides law enforcement and state governments with a **starting point** from which to begin addressing gang violence, intimidation and in our communities.

However, while many gangs today are easily identified by well publicized names, hand signs, and graffiti symbols, **it is important to recognize that a gang can actually exist even if some of the above listed elements are missing**. For example, in the emerging stages of a gang's formation cycle, a cohesive group of individuals can exist with loosely knit ties to a neighborhood or community, without a name, and be united for the common purposes of carrying out criminal activity, intimidation, drug activity and violence.

**Modern technology** has given informal and formalized gangs the ability to function in **complete anonymity online** using cell phones, instant messaging and the internet to communicate activities, goals and targets for future victimization, entirely unnoticed by local law enforcement agencies.

During a gang's early formation stages in some communities, any assumed name of the group may not be taken seriously by authorties, or may be seen as a high school joke because the members lack real and identifiable group clout, physical stature, or are not possessed of reputations for violent or intimidating behavior. Nevertheless, a gang may exist whether or not the police, local school authorities or even the local community openly validates it's activities, or has taken notice of the group's social transgressions in any publicly confirmatory and traceable manner.

Typically, affluent, suburban, and bedroom communities try to avoid the validation of gang activity and go through an extended period of **denial of the existence of gangs in their community** in order to preserve their community's reputation for safety, security, and peacefulness, and to avoid arousing fear and concern in the local residents. However, much of the denial results from the lack of familiarity and inexperience of the local police in dealing with gang activity, that is often thought to be associated with big city crime.

Consequently, the community denial stage of gang acknowledgment provides prime opportunities for gangs to flourish and become firmly entrenched in unsuspecting middle class and upper middle class communities while the residents are unknowingly perceiving uncharacteristic community incidents of crime, not as gang activity, but as someone else's juvenile delinquency problem.

Today, the growth of the gangs can occur almost overnight under one or more of the following circumstances:

- while adults are too busy working to notice the changes in their own child's behavior

- when no one is seriously looking with a concerned eye at changes in the activities of the youth in their community such as group **"beat downs"**, "bullying" and **"school intimidation"**

- when no one seems to really care about large numbers of youth hanging out in the community, at parks and in shopping centers

- when the community **falsely believes** that the problem is one of **temporary juvenile delinquency that the police know about**, will properly handle as necessary, or that the **problem will eventually go away on its own as the kids grow up**

- when **the police are inexperienced** with proper gathering and use of gang intelligence for use in both prevention and prosecution of gang-related incidents

Unfortunately, by the time that a gang has acquired a recognizable name, consistent pattern of activity and recognizable symbols; it is too late to curtail the aggregate damage, violence, influence and fear that gangs create in schools and within the local community. The police and community at this juncture must now operate from a defensive law enforcement position instead of an offensive position in addressing the root causes of gang activity in their communities that could have effectively begun during the community's denial phase.

In recent years, according to the **Maryland Gang Information and Prevention Office,** the following gangs have been identified as operating in varying levels within the following Maryland communities:

# Maryland Gang Chart

| City/County | Identified Gangs | Number of Gangs |
|---|---|---|
| Alleghany | <ul><li>Blood, Crips</li><li>Rollin 90's Neighborhood Crips</li><li>South Side Soldiers</li><li>Queen City Kings</li><li>Cumberland Gangstas</li><li>Unconfirmed-MS 13</li></ul> | |
| Anne Arundel County | <ul><li>Hood Fellas</li><li>Shipley's Choice Kings</li></ul> | Estimated: 25 gangs with more than 75 members. |

| | · MS13 | |
|---|---|---|
| | | |

**Figure 21**

# Maryland Gang Chart

| City/County | Identified Gangs | Number of Gangs |
|---|---|---|
| Baltimore City | Insane Red Devils<br>Tree Top PIRU Bloods<br>Edmondson Avenue Bloods<br>Treetop PIRU Bloods<br>North and Braddish Zone<br>Purple City Dip Set Bloods<br>Veronica Avenue Bloods<br>Veronica Avenue Boys<br>721 Eastside Blood<br>Bangers<br>Mara Salvaltrucha (MS-13)<br>Hillside Boys | Estimated: 170 known criminal street gangs with over 1300 members.<br><br>Estimated: 150 known members of Motorcycle gangs<br><br>Estimated; 250 gang members in the Baltimore City Detention Center |
| Baltimore County | Crips<br>Bloods<br>Mara Salvatrucha (MS-13)<br><br>Outlaw Motorcycle Gang (OMG) | Estimated : 35 gangs<br><br>The total membership of OMG members in Baltimore County is approximately 75 in Whitemarsh and Dundalk areas.<br><br>Rivals: Pagans and Hells Angels |
| Calvert County | Hells Angels | Limited Transient Activity<br>Hells Angels, MS-13 |
| Caroline County | MS-13<br>Bloods<br>Crips<br>Black Guerilla Family<br>Sureno<br>Latin Kings<br><br>Pagans Outlaw Motorcycle | Estimated: 7 |

| | Gang | |
|---|---|---|
| Carroll County | Bloods<br>Crips | Estimated: 9 gangs with approximately 77 active members. |

Figure 22

# Maryland Gang Chart

| City/County | Identified Gangs | Number of Gangs |
|---|---|---|
| Cecil County | No Known | |
| Charles County | The Cut Boyz<br>MS-13, Bloods<br>Knock Out Kings<br>Black Guerilla Family<br>Iron Horsemen Outlaw Motorcycle Gangs<br>Phantoms Outlaw Motorcycle Gangs | Estimated: 14 |
| Dorchester County | Crips<br>Bloods<br>Black Mambo Kings<br>Black Guerilla Family<br>MS-13 . | Minimal Unconfirmed |
| Frederick County | Crips<br>Westside Crips<br><br>MS-13<br><br>Bloods<br>C.M.G.<br>Cobraz<br>T.A.S<br><br>PA Mob<br>B6<br>Amber Meadows Crew<br>Kash Money Gang<br>White Soxs.<br><br>Juice Crew (white high school)<br>Newport Crew (white | Estimated:<br>Bloods approximately 30+ members<br><br>Crips approximately 40+ members |

| City/County | Identified Gangs | Number of Gangs |
|---|---|---|
| | high school) | |
| Garrett County | No Known | |

Figure 23

# Maryland Gang Chart

| City/County | Identified Gangs | Number of Gangs |
|---|---|---|
| Harford County | Bloods<br>Crips<br>Hells Angels<br>Pagans, | Estimated:<br>Bloods approximately 200+ members<br><br>Crips approximately 90+ members |
| Howard County | MS-13 | Limited Data. Emerging Gang Activity |
| Kent County | No Known | |
| Montgomery County | MS-13<br>Street Thug Criminals (STC)<br>Vatos Locos (VL) | Estimated:<br>45 major gangs with total gang of 1000 largely concentrated in Silver Spring, Wheaton, Gaithersburg, and Germantown |
| Prince George's County | Bloods, Crips,<br><br>Mara Salvatrucha (MS-13)<br>Street Thug Criminals (STC)<br>Vatos Locos (VL)<br>Bloods<br>Crips | Estimated: 400+ members.<br><br>Major gangs concentrated on county borders of District of Columbia and Montgomery County in Langley Park, Lanham, Oxon Hill, and Suitland.<br><br>Estimated: 50 "crews" or African American gangs |
| Queen Anne's | Unverifiable | |
| Somerset County | Hot Boys<br>MS-13<br>Dog Town Posse | Estimated:3 |

Figure 24

84

# Maryland Gang Chart

| City/County | Identified Gangs | Number of Gangs |
|---|---|---|
| St. Mary's County | MOB | Limited Activity & No Related Crime |
| Washington County | Bloods<br>Crips<br>Westside Crips<br>Money Making Crips<br>Outlaw Gangsters Crips<br>61st North Hoover Crips<br>MS-13 | Estimated:<br>Bloods 100<br>Crips 75<br><br>primarily in the city of Hagerstown. |
| Worcester County | Hells Angels,<br>Pagans<br>MS-13 | Transient. No Gang Related Crime |
| Wicomico County | East Side<br>FTL<br>ABM (All About Money)<br>MG (Murder Gang | Estimated: 4<br>Primarily in Salisbury and Fruitland |

Source: Charts 21-25 http://gangs.umd.edu/wfrmByLocation.aspx

**Figure 25**

This list of gangs is not intended as an exhaustive representation of the total gang activity that may actually be occurring in each Maryland jurisdiction.

## Maryland Exile – Attacking Violent Crime Today

**Maryland Exile**: **Maryland Exile** is a multi-jurisdictional approach to combating violent gang activity in the 21st century in Baltimore, Maryland utilizing the collaborative resources of the Baltimore City State's Attorney's Office, local police and the Baltimore U.S. Attorneys Office.

**Prosecutors in the Baltimore City State's Attorneys' Office** working in partnership with the **Baltimore U.S. Attorneys Office** have taken an aggressive stance on gangs and gun-related crimes. Under the auspices of the Maryland Exile initiative at both the state and federal levels, prosecutors are targeting the complete dismantling of violent gangs and organizations who use guns, drugs and violence as their stock in trade in Baltimore City, Prince Georges' County and Wicomico County.

At the core of the Maryland Exile program exists an iron-fisted crime fighting strategy to rapidly identify, prosecute, communicate with, and jail all persons in either state or federal prison who meet the basic program gun violation criteria. The current **"use a gun equals go to jail"** philosophy is being implemented at all levels within the criminal justice system in Maryland from the police to prosecutors, probation and parole agents to corrections officials.

The Maryland Exile program effectively means that there are no longer any probation plea bargain deals being cut for gun related crimes in targeted exile areas. Doing **"hard time"** is the rule of the day!

Source: http://www.usdoj.gov/usao/md/Exile/files/Baltimore%20EXILE%20strategy%20final.2006.02.15.pdf

## Maryland Gang Prosecution Act

In addition to the implementation of the Maryland Exile initiative, Maryland passed the **Maryland Gang Prosecution Act of 2007** as one more tool to fight the growing statewide gang problem.

The Maryland Gang Prosecution Act of 2007 prohibits a person from:

- participating in a certain criminal gang knowing that the members of the gang engage in a specified pattern of criminal gang activity
- prohibits a person from knowingly and willfully directing or participating in a specified criminal offense committed for the benefit of, at the direction of, or in association with a criminal gang
- prohibits a person from committing a violation of the Act involving the commission of an offense that results in the death of the victim.

According to Maryland Senator Roy Dyson in an online article, *"Opinion: Legislation to Prevent Gang Activity Passes in Annapolis"*, the Maryland Gang Prosecution Act of 2007 was modeled after the federal Racketeering Influenced and Corrupt Organization Act (RICO) and violators of this law would be guilty of a felony.... be subject to 30 years imprisonment.... and up to a $300,000.00 fine. (Emphasis added)

# CHAPTER TEN

# COMPUTER CRIME AND COMPUTER FORENSICS

SECTION 10: COMPUTER CRIME AND COMPUTER FORENSICS

In this chapter we will explore some general of computer forensics principles to increase each student's familiarity with important basic computer forensics concepts, and some rules related to conducting sound computer forensics examinations for hidden electronic evidence.

As a practical matter, computers and electronic data have become an integral part of our daily lives. They control virtually every aspect of our existence from the operation of vehicles, to the identification of each cell phone call that we make. Computers also control many household items such as refrigerators, televisions, lights, surveillance and alarm systems and personal electronic devices such as Ipods, watches, mp3 players, digital cameras, thumb drives, and gaming devices. Consequently, all of these electronic devices provide vast opportunities to hide information related to criminal activity, or to gain unauthorized access to confidential personal and financial data for use during the commission of a computer-related crime.

**Computer Crime/Cybercrime** – Computer Forensics has more recently become known as Digital Forensics and is one of the fastest growing disciplines in the criminal justice field. Computer/Digital Forensics combines the principles of criminal investigation, law, the rules of evidence, and information technology for use in detecting illegal activity in the areas of child pornography, identity theft, unauthorized computer access, theft of trade secrets, drug crimes, murder, fraud, financial crimes and other electronic criminal offenses.

**Forensics Defined:**

Generally, **Forensics** is defined as:
- the application of scientific knowledge to legal issues or problems; or
- the application of scientific techniques for investigating, preserving and examining evidence in a particular field to establish an evidentiary basis for use in court cases

## Computer Forensics Defined:

Generally, **Computer Forensics** :

- is the application of forensic techniques to electronic information stored or transported on computers

- involves the preservation, identification, extraction, examination, documentation, and interpretation of computer media for evidentiary use in legal proceedings, administrative hearings, and business.

Although several variations exist today for the definition of computer forensics, they all have a common purpose of applying forensic computer investigative techniques and analysis to gather evidence suitable for presentation in a court of law or business matter.

Given the limitless number of electronic storage media devices in the marketplace today, the term "digital forensics" is being used with greater frequency to cover all electronic storage media devices, including computers. At the present, the term computer forensics has become interchangeable with digital forensics.

## Computer Crimes

Computer crimes generally fall into two categories:

- where the **computer** is **used** to commit the crime such as in child pornography cases, to send threatening letters, or to commit fraud; or

- where the **computer** is the **target** of the crime such as in unauthorized access cases, curiosity cases, or in hacking cases resulting in stolen data or system damage

Computer crimes are commonly called cyber crime in many jurisdictions.

## The "Triple A" Approach to Computer Forensics Examination

The **"Triple A"** Approach to computer forensics examination refers to the process
of :

- Acquiring
- Authenticating, and;
- Analyzing electronic evidence

The Triple A Approach to computer forensics evidence is used to ensure that the original evidence media (i.e. hard drive) remains unaltered and free from modification by the computer forensics examiner to retain its reliability. The absence of any modification of the media makes the computer examination results more likely to be admitted into evidence in a court of law, a government proceeding or in a business matter. The admissibility of the computer forensics evidence will also be dependent upon whether the computer forensics examiner is determined by the courts (or another responsible official) to be qualified as either a computer forensics technician or as a computer forensics expert witness based upon his overall education, training and experience.

## Role of the Computer Forensics Examiner

A well trained computer forensics examiner using specialized computer forensics software tools is normally required to:

- **acquire** the electronic evidence without damaging the original data,
- **authenticate** or verify that the recovered evidence is the same as the originally seized data located on a suspect's computer
- **analyze** the data without modifying it

Most law enforcement agencies have at least one computer forensics examiner on staff, or available to assist them with cybercrime offenses. However, the sheer number of child pornography cases occurring throughout the U.S. today may account for approximately 70% of all law enforcement computer crime cases, and is overwhelming the resources and personnel in many local law enforcement agencies.

In Maryland, a limited number of criminal justice professionals within various government agencies have been trained in some form of computer forensics examination analysis in the Maryland Attorney Generals' Office, the Maryland Department of Public Safety, the Maryland State Police and the Maryland State Prosecutor's Office.

For example, a Maryland Division of Corrections parole agent or a supervising community correctional officer may have been trained to conduct computer forensics examinations on computers owned by parolees released back into the community to help confirm that the parolee is not engaged in any illegal conduct, has not illegally accessed a computer or has not possessed any illegal images in violation of their parole supervision conditions.

Many computer forensics examiners typically have training using two or more of the following computer forensics software tools:

- NTI Tools (Dos Tools)
- Encase
- Forensic Tool Kit (FTK)
- Norton Utilities Diskedit
- Paraben (Cell Phones, PDA's & Hand held devices)
- X-Ways (Win Hex)

- Various Open Source Tools (Helix, Snort, etc.)
- Pro Discover
- Miscellaneous additional computer forensics tools

## Computer Forensics Careers

Computer forensics is also one of the fastest growing employment areas in the country. Individuals with specialized training in the field are likely to find greater employment opportunities in the future. Additionally, well trained individuals may gain a distinctive advantage in the workplace as the need increases for companies, agencies and police departments to properly seize, preserve, and examine electronic computer evidence related to cybercrime investigations and electronic discovery.

According to a Bureau of Labor Statistics report cited in a September 1, 2006 article "Ten Hot Jobs for 2007", the job of Computer Forensic Expert was listed as one of the cutting edge career choices for 2007 (Comer, 2006).

# CHAPTER ELEVEN
# CRIMINAL JUSTICE ON
# MARYLAND'S EASTERN SHORE

SECTION 11: CRIMINAL JUSTICE ON
MARYLAND'S EASTERN SHORE

## The Eastern Shore Criminal Justice System

**Maryland's Easter Shore** – Maryland's Eastern Shore is a vastly diverse, rural region of Maryland rich in criminal justice history.  Despite the occurrence of high profile criminal cases from time to time, the Eastern Shore criminal justice system has escaped recognition and review in modern criminal justice literature. Nevertheless, the Eastern Shore is experiencing unprecedented population growth, tourism expansion, increased industry relocations and a housing boom that will bring with it many of the same criminal justice issues that are occurring daily in other Maryland jurisdictions. Consequently, this chapter briefly takes a look at the characteristics of the Eastern Shore criminal justice system and introduces the criminal justice student to some unique environmental crimes that are likely to occur in the agricultural, recreational and waterfront communities of the Eastern Shore.

Emphasis in this chapter is placed on the key role played by the Maryland Department of Natural Resources Police (DNR), whose enforcement efforts have been vital to the preservation of Mary's tremendous natural resources.

## Eastern Shore Courts

The following Court Chart illustrates the characteristics of the Maryland Eastern Shore Courthouses, the population size served in each county, and the number of judges responsible for administering the community's justice:

## Eastern Shore Courthouses

| Court Location | Number of Judges | 2003 Population* |
|---|---|---|
|  |  |  |
| Caroline County |  | 30,861 |
| District Court | 1 |  |
| Circuit Court | 1 |  |
|  |  |  |
| Calvert County |  | 84,110 |
| District Court | 1 |  |
| Circuit Court | 2 |  |

## Eastern Shore Courthouses

| Court Location | Number of Judges | 2003 Population* |
|---|---|---|
| Cecil County | | 92,746 |
| District Court | 2 | |
| Circuit Court | 3 | |
| | | |
| Charles County | | 133,049 |
| District Court | 2 | |
| Circuit Court | 4 | |
| | | |
| Dorchester County | | 30,612 |
| District Court | 1 | |
| Circuit Court | 1 | |
| | | |
| Somerset | | 25,447 |
| District Court | 1 | |
| Circuit Court | 1 | |
| | | |
| Talbot County | | 34,670 |
| District Court | 1 | |
| Circuit Court | 1 | |
| | | |
| Wicomico County | | 87,375 |
| District Court | 2 | |
| Circuit Court | 3 | |
| | | |
| Worcester County | | 49,604 |
| District Court | 1 | |
| Circuit Court | 2 | |
| | | |
| Queen Anne's County | | 44,108 |
| District Court | 1 | |
| Circuit Court | 1 | |
| | | |
| St. Mary's County | | 92,754 |
| District Court | 1 | |
| Circuit Court | 3 | |

**Source:** U.S. Census Bureau, ePodunk Population Growth of Counties 2003

**Figure 26**

# Eastern Shore Criminal Offenses – On the Water

Some Typical Eastern Shore criminal offenses related to natural resources include:

## Water- Related Offenses:

- Failure to Mark Equipment
- Dredging in a Restricted Area
- Failure to Have Fishing License for Inspection
- Disorderly Conduct
- Setting an Unattended Gill Net
- Possession of Unculled Oysters
- Possession of Undersized Hard Crabs
- Possession of Undersize Oysters
- Failure to Complete & Sign Dockside Vouchers
- Setting Excess Crab Pots in Violation of Maryland law
- Making False Entries into a Public Document (fishing reports)
- Failing to Transmit Accurate Records as Required by DNR
- Operating a Vessel Under Influence of Alcohol
- Selling Undersized Chesapeake Bay crabs interstate (a felony under the federal Lacey Act)

## Non Water – Related Criminal Offenses

- Possession of Marijuana
- Underage Possession of Alcohol
- Driving While Intoxicated
- Possession of Controlled Dangerous Substances & Paraphernalia

Many of the criminal offenses listed is this section are likely to occur at popular fishing locations on the Chesapeake Bay, in Ocean City, Maryland, on campgrounds, and in various communities adjacent to Maryland waterways, inlets and state parks.

# CHAPTER TWELVE

# MARYLAND NATURAL RESOURCES ENVIRONMENTAL LAW RESEARCH SOURCES

## SECTION 12: MARYLAND NATURAL RESOURCES RESEARCH SOURCES

### Maryland Natural Resources - Research Sources

The following websites contain several Maryland Natural Resources environmental laws and regulations affecting the regions of the Eastern Shore in which criminal offenses are likely to occur in fishing locations, on hunting grounds, aboard recreational boats, and in state parks on the bay.

♦ Summaries of Maryland fishing rules and regulations
http://www.dnr.state.md.us/fisheries/regulations/regindex.html

♦ Maryland State & Federal requirements for recreational vessels
http://www.dnr.state.md.us/boating/regulations/

♦ Proposed boating regulations for 2004
http://www.dnr.state.md.us/boating/regs.html

♦ Hunting Regulations
http://www.dnr.state.md.us/huntersguide/index.html

♦ Rivers and Streams: Restoring and Protection Regulations
http://www.dnr.state.md.us/streams/res_protect/regulations.html

♦ Bay Grass Regulations
http://www.dnr.state.md.us/bay/sav/regulations.html

♦ State Park Regulations
http://www.dnr.state.md.us/publiclands/dccomarsp.html

♦ Roadside Tree Care Regulations
http://www.dnr.state.md.us/download/rsregs.pdf

♦ Forest Management Open Air Burning Regulations
http://www.dnr.state.md.us/forests/otheragencies/firenotes.html

# Chapter Thirteen

# Notable Maryland
# Criminal Offenses

This section was designed to give criminal justice students a more in-depth review of Maryland criminal justice laws that may be applicable to college student life today because of the growing frequency of reported occurrences of excessive alcohol consumption, drug use and sexual activity on college campus communities throughout the state.

The following Notable Offenses represent timely, informative and useful information for the Maryland criminal justice student because of the increased media coverage of college student behaviors related to assaultive-type conduct, alcohol offenses, sexual oriented conduct, stalking, and drug offenses.

Our expectation is that students will develop a proactive awareness of how these laws might apply to their everyday conduct or the conduct of someone that they may know. A secondary consideration is to promote the avoidance of any type of involvement by students in the offenses listed below who are seeking a future career in the criminal justice field. Police departments today are closely scrutinizing binge drinking by college students, recent drug use within five years prior to the application for a law enforcement position, and assaultive conduct.

Students can also research college campus crime rates at:
- http://ope.ed.gov/security/
- http://www.securityoncampus.org/crimestats/index.html

College crime rates are available on every campus that receives federal funding and students should be able to obtain annual crime data from the main web page of any college in which they are interested.

# Notable Offenses

The following Maryland offenses have been presented accurately but in an abbreviated form by eliminating much of the legalese connected with each crime. Students should conduct an independent review of the original sources cited for each offense to learn more about the inclusive and excepted conduct related to each listed criminal statute.

## Assaultive Crimes

.

**Assault 1st Degree** – First Degree Assault involves the use of a firearm to commit an assault, or intent to cause serious physical injury during the commission of an assault.

**Punishment:** Felony – up to 25 years imprisonment

Source: Md. CRIMINAL LAW Code Ann. § 3-202 (2007)

---

**Assault 2nd Degree** – Second Degree Assault involves the intentional frightening of another person, without legal justification, by threatening offensive or harmful contact, coupled with the apparent ability to cause the contact or harm, and causing the victim to reasonably believe that he would immediately be harmed.

**Punishment:** Misdemeanor – up to 10 years imprisonment, a $2,500 fine or both

If victim is a Police Officer - crime is a Felony with punishment up to 10 years, a
$10,000 fine or both

Source: Md. CRIMINAL LAW Code Ann. § 3-203 (2007)

## Sex Offenses

**1st Degree Sex Offense (Rape)** – First Degree Sex Offense involves the unlawful sexual intercourse with a female by use of force, fear, or the threat of force, without consent, and with penetration however slight. Force includes use of a dangerous weapon, suffocation, serious physical injury, disfigurement, kidnapping or aiding & abetting another individual or during commission of a burglary.

**Punishment:** Felony – up to Life Imprisonment

Source: Md. CRIMINAL LAW Code Ann. § 3-303 (2007)

---

**2nd Degree Sex Offense (Rape)** – Second Degree Sex Offense involves the unlawful sexual intercourse with a female, by use of force, fear, or the threat of force,
without consent, and with penetration however slight.

**Punishment:**    Felony – up to 20 years imprisonment

Source: Md. CRIMINAL LAW Code Ann. § 3-304 (2007)

---

**2nd Degree Sex Offense (Oral/Anal)**  – Second Degree Sex Offense (Oral\Anal) can also involve oral sex (fellatio/cunnilingus) or anal sex by use of force, fear or threat of force, without consent, wherein the mouth of a male or female is placed on the genital of the another, or the penis of a male is inserted into the anus of another.

**Punishment:**    Felony – up to 20 years imprisonment

Source: Md. CRIMINAL LAW Code Ann. § 3-306 (2007)

---

**2nd  Degree Sex Offense (Age Difference)** – Second Degree Sex Offense (Age) also involves oral sex (fellatio/cunnilingus) or anal sex with a person who was under the age of 14 years at the time, and the person committing the act [defendant] is at least 4 years older that the victim.

**Computation:** A basic computation means that the victim:

- is 14 years old and the defendant is at least 18 years old.

**Punishment:** Felony – up to 20 years imprisonment

Source: Md. CRIMINAL LAW Code Ann. § 3-306(a)(3) (2007)

---

**2nd Degree Sex Offense (Unconscious/Mental Defect)** – Second Degree Sex Offense (Unconscious\Mental Defect) also involves oral sex (fellatio/cunnilingus) or anal sex with a person who was unconscious at the time, helpless or suffers from a mental defect (retarded) that prevented the giving of voluntary consent to the sexual act(s).

**Application:** This statute would also seem to apply to persons who are heavily intoxicated or under the influence of drugs or are passed out at a party or during a date.

**Punishment:** Felony – up to 20 years imprisonment

Source: Md. CRIMINAL LAW Code Ann. § 3-306(a)(2) (2007)

---

**3rd Degree Sex Offense (Sexual Contact)** – Third Degree Sex Offense involves the intentional touching of the genital area, anal area or the intimate parts ("sexual contact") of another for the purpose of sexual gratification, sexual arousal or for abuse by using any part other than the penis, mouth or tongue, against the will of victim and without the victim's consent under one or more of the following circumstances: (1) use of a weapon, infliction of violence on the victim or anyone else during the commission of the offense (2) while aid and abetting another during the commission of the offense, (3) or during the commission of a burglary, but does not include common acts of friendly expressions, affection or for medical purposes.

**Application:** This offense can occur with the use of the hand, finger [or possibly with a foreign object].

**Punishment:** Felony – up to 10 years imprisonment

Source: Md. CRIMINAL LAW Code Ann. § 3-307(a)(3) (2007)

---

**3rd Degree Sex Offense (Under Age 14)** – Third Degree Sex Offense (Age) involves the intentional touching of the genital area, anal area or the intimate parts ("sexual contact") of another for the purpose of sexual gratification, sexual arousal or for abuse by using any part other than the penis, mouth or tongue, against the will of victim and without the victim's consent under one or more of the following circumstances: (1) use of a weapon, infliction of violence on the victim or anyone else during the commission of the offense (2) while aid and abetting another during the commission of the offense, (3) or during the commission of a burglary, of any person who was under 14 years old at the time of the contact and the defendant is at least 4 years older than the victim. Sexual contact does not include common acts of friendly expressions, affection or for medical purposes.

**Computation:** A basic computation means that the victim is :

- at least **13 years old** and the defendant is at least **17 years old**

**Punishment:** Felony – up to 10 years imprisonment

Source: Md. CRIMINAL LAW Code Ann. § 3-307(a)(3) (2007)

---

**3rd Degree Sex Offense (Intercourse -Age)**–Third Degree Sex Offense (Intercourse-Age) involves vaginal intercourse with a victim who was either 14 or 15 years old at the time of the intercourse, and the defendant was 21 years at the time of the sexual intercourse.

**Computation:** A basic computation means that the victim is:

- at least **14 years** old or **15 years** old, and ;
- the **defendant is 21** years old having sex with a 14 or 15 year old                                                        victim.

This is a **strict liability offense** meaning that "voluntary consent" by the victim is not a defense to the crime. If the sexual intercourse is proven as having been performed, the guilt of the defendant is established as a matter of law. Intent of the defendant is not an element of the crime, meaning that it does not matter what the defendant intended to do at the time. This offense is commonly known as "statutory rape" in many states.

**Punishment:** Felony – up to 10 years imprisonment

Source: Md. CRIMINAL LAW Code Ann. § 3-307(a)(4) (2007)

---

**4th Degree Sex Offense (Sexual Contact)** – Fourth Degree Sex Offense involves the intentional touching of the genital area, anal area or the intimate parts ("sexual contact") of another for the purpose of sexual gratification, sexual arousal, or abuse without consent, by using any part other than the penis, mouth or tongue, but does not include common acts of friendly expressions,        affection        or        for        medical        purposes.

Fourth Degree Sex Offense (Sexual Contact) contains all the elements of third degree sex offense except the <u>aggravating circumstances</u> of i.e. (1) use of a weapon, infliction of violence on the victim or anyone else during the commission of the offense (2) while aid and abetting another during the commission of the offense, (3) or during the commission of a burglary.

**Application:** This offense can occur with the use of the hand, finger or possibly with a foreign object. Identifiable examples may include unwanted touching of breasts, grabbing someone's buttocks in tight jeans, feeling on another's thighs or crotch , pulling down a female or male's gym pants or sweat suit to expose their underwear, or unwanted kissing on the cheek, etc.

**Punishment:** Misdemeanor – up to 1 year imprisonment or a $1,000 fine or both.

Source: Md. CRIMINAL LAW Code Ann. § 3-308(a)(1) (2007)

---

**4th Degree Sex Offense (Sexual Act-Age)** – Fourth Degree Sex Offense (Age) involves oral sex (fellatio/cunnilingus) or anal sex ("sexual act") with a victim who was either 14 or 15 years old at the time of the contact, and the defendant was at
least 4 years older than the victim at the time  of the sexual contact, or the defendant was 21 years old.

**Computation:**  A basic computation means that the victim is:
- at least 14 years old or 15 years old  and
- the defendant is at least either 18 years old involved with a 14 year old victim, or;
- the defendant is either 19 years old involved with a 15 year old victim, or;

This is a **strict liability offense** meaning that "voluntary consent" by the victim is not a defense to the crime. If the sexual act is proven as having been performed the guilt of the defendant is established as a matter of law. Intent of the defendant is not an element of the crime. This offense does not include intercourse with the victim.

**Punishment:** Misdemeanor – up to 1 year imprisonment or a $1,000 fine or both.

Source: Md. CRIMINAL LAW Code Ann. **§ 3-308 (a)(2)(3) (2007)**

# HOMICIDE CRIMES

## MANSLAUGHTER - TWO TYPES

**Voluntary Manslaughter** – **Voluntary Manslaughter** involves the intentional killing of another where:
- the defendant acted with a "hot blooded response" to legally adequate provocation, which resulted in a rage that had not cooled at the time of the killing; or
- where the defendant used more force than was necessary in defending himself (partial/imperfect self defense); or
- used more force than was necessary in defending a third party (partial/imperfect self defense); or
- where the defendant kills someone while under duress (force) that he would be imminently killed or suffer serious bodily harm and with no reasonable opportunity to escape

**Involuntary Manslaughter** –**Involuntary Manslaughter** involves conduct by the defendant:

- that caused the death of another, and the conduct was "grossly negligent" in that it created a high degree of risk to human life; or
- the defendant committed an "unlawful act" during which someone was killed by the defendant or another person, and the unlawful act resulted in the death of the victim

Maryland makes no distinction between voluntary and involuntary manslaughter for purposes of punishment in that both offenses carry the same sentence.

**Punishment:** Felony – up to 10 years imprisonment, or imprisonment in a local

correctional facility up to 2 years or a $500.00 fine or both.

Spousal

adultery involving discovery of one's spouse having sexual intercourse with another person is no longer a mitigating factor

to

reduce murder to manslaughter as previously existed under

the

Maryland common law manslaughter.

Source: Md. CRIMINAL LAW Code Ann. § 2-207 (a) (b) (2007)

## Drug Crimes

Illegal drug use and sales are a problem throughout the State of Maryland and the penalties for Maryland drug violations reflect a historically hard-line approach to drug crime. However, in recent years, the Maryland Drug Courts have emerged to take the lead in dealing with the voluminous drug offense court dockets by taking a medical treatment approach to combating the increasing statewide drug problem.

### Offenses

**Possession of a Controlled Dangerous Substance (CDS)** [including Marijuana] – involves the knowing possession of an illegal substance with either actual control or constructive/indirect possession (i.e. constructive control together with someone else) where the surrounding circumstances indicate ownership of the controlled dangerous substance.

It is illegal to possess or administer specified controlled dangerous substances to another, [or to give it away] or to obtain or attempt to obtain controlled dangerous substances by fraud, deceit, concealment, misrepresentation, or use of a false name or counterfeit prescription.

**Punishment:**   All offenses are Misdemeanors
Punishment:   For CDS – up to 4 years imprisonment, or a $25,000.00 fine or both
Punishment:   If Marijuana – up to 1 year imprisonment, or a $1,000.00 fine or both
Punishment:   If Marijuana with Medical Necessity – up to $500.00 fine

Source: Md. CRIMINAL LAW Code Ann. § 5-601(a) (2007)

## Theft-Type Crimes

**Fourth Degree Burglary (Vehicle)** - *Rogue and Vagabond* – involves the possession of burglary tools with the intent to use them to commit the breaking and entering of a motor vehicle, or the presence in another's vehicle with intent to commit a theft of the vehicle or of property inside of the vehicle.

**Punishment:**   Misdemeanor – up to 3 years imprisonment & person is deemed                                                                                                   a
rogue and vagabond under Maryland Law

Source: Md. CRIMINAL LAW Code Ann. § 6-206 (2007)

## Maryland Consolidated Theft Statute Crimes:

In Maryland, theft can be committed in one or more of the following ways:

- Knowingly or willfully obtaining or exerting **Unauthorized Control** over property
- Obtaining control of property by **Deception**
- Possession of **Stolen Property**
- Inference of Theft drawn from exclusive **unexplained possession** of **Recently Stolen Property**

[§ 7-102(a) Theft –Type Crimes]

- Failure to pay for **Services** known to be provided only for compensation
- Larceny by **Trick**
- Embezzlement
- False Pretenses
- Shoplifting

Punishment:    Felony Theft Over $500.00 – up to 15 years imprisonment, or $25,000.00 fine or both

Punishment:    Misdemeanor Theft Under $500.00 – up to 18 months imprisonment,
      or $500.00 fine or both

Punishment:    Misdemeanor Theft Under $100.00 – up to 90 days imprisonment, or
      $500.00 fine or both

Source: Md. CRIMINAL LAW Code Ann. § 7-104 (a)(b)(c)(d) (2007) &
      Md. CRIMINAL LAW Code Ann. §§ 7-102(a) (2007)

---

## Maryland Computer Crime:

---

**Unauthorized Access to Computers** – involves:

- the intentional, willful, and unauthorized access,
- attempts to access or
- exceeding the scope of authorization access for all or a part of a computer network, computer system, database or related electronic media and software, and

- prohibits acts intended to interrupt, damage, destroy a computer system, identify the access code, or publicize the access code to an unauthorized person.

Punishment: Misdemeanor – Generally up to 3 years imprisonment, or $1000.00 fine or both

Punishment: Felony if loss is $10,000.00 or more – up to 10 years imprisonment, or
$10,000.00 fine or both

Punishment: Misdemeanor if loss is less than $5,000.00 – up to 5 years imprisonment, or both

Source: Md. CRIMINAL LAW Code Ann. § 7-302 (2007)

## Handgun Violations:

**Wearing, Carrying or Transporting Handgun** – involves the wearing, carrying or transporting of a handgun (pistol, revolver) or firearm capable of being concealed on or about the person that is designed to fire a bullet by the explosion of gunpowder. The handgun can be concealed or carried openly and can be either in reach or available for the person's use.

**Punishment:  Misdemeanor** – 30 days to 3 years imprisonment, and fine of $2500.00 to $2,500.00 or both

**Punishment: If Prior Convictions** – Sentence can range from 1 year to 10 years.

Source: Md. CRIMINAL LAW Code Ann. § 4-203 (2007)

## Alcohol Citations: (Civil Offense)

**Alcohol Citation** (Possession Alcohol Under 21) – An alcohol citation is **a civil offense**, not a criminal offense and involves the possession of alcohol by a person under the age of 21 years. Alcohol citations are handled and processed through court system in the same manner as criminal offenses in the State of Maryland in that police officers can issue citations, a court date is scheduled and Prosecutors handle the cases on the court docket.  The court hears evidence and decides if the person is guilty of a "civil offense" only, not a criminal offense.

**Punishment:** Civil Offense – $500.00 or $1,000.00 fine. Most violators cited agree to attend a diversion program and have their cases stetted or nolle prosequi'd after completion of diversion. There is no criminal record. However, a judge

may consider the alcohol incident in future criminal case dispositions.

Source: Md. CRIMINAL LAW Code Ann. §10-114 (2007)

---

## Relationship-Type Personal Crimes:

**Stalking:** – Stalking is a malicious (intentional or willful) course of conduct intended, or should reasonably be known to place a person in fear of serious bodily injury, assault, rape or sexual offense, false imprisonment, or death.

**Punishment:** Misdemeanor up to 5yrs imprisonment, and fine of $5,000.0 or both

Application: Stalking frequently occurs when dating relationships, marriages and certain friendships breakdown, become embattled, or may result from other personal conflicts, neighbor disputes or business dealings. At other times stalking results from unwanted attention given to victims by persons unknown to the victim.

Source: Md. CRIMINAL LAW Code Ann. § 3-802 (2007)

---

**Telephone Misuse:** – Telephone misuse involves using telephone equipment to make anonymous calls that are reasonably expected to annoy, abuse, harass or embarrass a person or making repeated calls to also torment the recipient, or to make proposals, suggestions, comments or requests that are obscene, lewd, lascivious, filthy or indecent.

**Punishment:** Misdemeanor – up to 3yrs imprisonment, and fine of $500.00 or both

Application: Telephone Misuse can occur with the making of repeated prank calls, and can also occur when dating relationships, marriages, and certain friendships breakdown or become embattled. Other times telephone misuse can result from unwanted attention given to others by persons unknown to the victim.

Source: Md. CRIMINAL LAW Code Ann. § 3-804 (2007)

---

**Misuse of Electronic Mail (Email):** – Misuse of Electronic Mail (Email) involves the use of email to harass one or more persons, or by sending lewd, lascivious or obscene material.

**Punishment:**     Misdemeanor – up to 1yrs imprisonment, and fine of $5,000.00                                                                                   or
          both.

Application: Misuse of Electronic Mail frequently occurs like stalking, and telephone misuse when dating relationships, marriages, roommates, and certain friendships breakdown or become embattled, particularly since multiple communication devices can be contained in a user's single electronic unit such as a cell phone, PDA, blackberry, or a laptop computer, etc.

 Source: Md. CRIMINAL LAW Code Ann. § 3-805 (2007)

---

**Harassment:** – Harassment involves the following a person around in public or engaging in a course of conduct that alarms or seriously annoys or harasses another without legal purpose after receiving a request to stop.

**Punishment:**     Misdemeanor – up to 90 days imprisonment, and fine of $5,000.00                                                                                   or
          both.

Application: Harassment can come in many forms and can also be connected with
other relationship-type crimes or from incidents that occur in daily life such as a car accident, or living next to a bad neighbor, etc.

Source: Md. CRIMINAL LAW Code Ann. § 3-803 (2007)

---

# CHAPTER FOURTEEN
# NOTABLE MARYLAND CASE DECISIONS

SECTION 14: NOTABLE MARYLAND CASE DECISIONS

## Notable Maryland Case Decisions for Discussion

This case law discussion section is designed to stimulate thought and discussion on some interesting, unusual or noteworthy criminal cases decided from time to time by the Maryland Courts. Each case can be used as an individual assignment or group assignment in class to elicit student reactions to the court decision, or can serve as an independent research assignment project.

The cases contained in this section will expand over the life time of this Supplement.

**Maryland Court Cases:**

### Identity Theft & Assuming A Fictitious Identity

Kazeem Adeshina Ishola v. State of Maryland – During trial evidence was presented that Ishola had opened an account using a name other than his own and had attempted to open one or more accounts using a fictitious name. The conviction was affirmed by the Court of Special Appeals and the court held that a person is guilty of identity fraud/theft if they attempt to use a fictitious name other than their own.

The case was appealed to the Court of Appeals September 2007. Kazeem Adeshina Ishola v. State of Maryland, 175 Md. App. 201 (2007).

Application: The Md. CRIMINAL LAW Code Ann. §8-301( c) Identity Fraud law in this case requires proof that a person assumed another's identity for purposes to avoid identification, prosecution or with fraudulent intent to obtain a benefit, service, goods, credit, etc, or to avoid payment of a debt.

Another related law is Md. CRIMINAL LAW Code Ann. § 10-115 (fake identification) which may apply to cases involving the use of a fake identity or a fake driver's license by underage teens or college students to purchase liquor or to get into night clubs.

What are your thoughts on this case?

_____


**Police Search of a Suspect's Buttocks for Drugs – In Public View**

John August Paulino v. State of Maryland – Paulino was a vehicle passenger and the police, through an informant were notified that Paulino was carrying drugs in his buttocks. Paulino's pants were worn low (sagging). Once the vehicle was stopped, Paulino was identified and police searched inside Paulino's buttocks while wearing gloves, in the public view of others and recovered drugs. The case was ultimately reversed due to lack of exigent (emergency) circumstances and the highly intrusive nature of the search in violation of the Fourth Amendment to the Constitution.
John August Paulino v. State of Maryland, 399 Md. 341 (2007)

What are your thoughts on this case?

_____

# APPENDIX FORMS
## Charging Document –Application for Statement of Charges

**DISTRICT COURT OF MARYLAND FOR** _____ (City/County)

LOCATED AT (COURT ADDRESS)

_____

_____

DISTRICT COURT CASE NUMBER

RELATED CASES:

_____

_____

_____

COMPLAINANT/APPLICANT

DEFENDANT

Printed Name

Printed Name

Number and Street Address

Number and Street Address

City, State, and Zip Code          Telephone

City, State, and Zip Code          Telephone

CC#

Agency, sub-agency, and I.D. #          (Officer Only)

DEFENDANT'S DESCRIPTION: Driver's License# _____ Sex ____ Race ____ Ht _____ Wt _____

Hair _____ Eyes _____ Complexion _____ Other _____ D.O.B _____ ID _____

### APPLICATION FOR STATEMENT OF CHARGES FOR BAD CHECK

I, the undersigned, apply for statement of charges or warrant which may lead to the arrest of the above named

Defendant because on or about _____ at _____
                                    Date                         Place

_____ , the above named Defendant

did unlawfully obtain _____ having a value of $_____

from (full legal name of business or person): _____

by uttering a certain bad check dated: _____ Check No: _____

ACCOUNT NO: _____ Drawn by: _____

on the (name/address of bank): _____

presented to (full legal name of business or person): _____ Payable to: _____

Said check was returned from bank marked: _____ on (date) _____

REGISTERED LETTER SENT (date): _____ RETURNED MARKED: _____

(Continued on attached _____ pages) (DC/CR 44A)

I solemnly affirm under the penalties of perjury that the contents of this Application are true to the best of my knowledge, information and belief.

_____          _____
          Date                           Officer's Signature

I have read or had read to me and I understand the Notice on the back of this form.

_____          _____
          Date                           Applicant's Signature

Subscribed and sworn to before me this _____ day of _____ , _____
                                                              Month              Year

Time: _____ M          Judge/Commissioner _____ I.D. _____

I understand that a charging document will be issued and that I must appear for trial ☐ on _____
                                                                                          Date
at _____ . ☐ when notified by the Clerk, at the court location shown at the top of this form.
        Time

☐ I have advised applicant of shielding right.   ☐ Applicant declines shielding.

☐ I declined to issue a charging document because of lack of probable cause.

_____          _____
          Date                           Applicant's Signature

                                 _____
                                 Commissioner          I.D.

TRACKING NUMBER

**DC/CR 44** (Rev. 12/2006)

# NOTICE TO APPLICANT FOR A CHARGING DOCUMENT

You are making an application for a charging document which may lead to the arrest and detention of the individual you are charging. If, as result of your application, a charging document is issued by the commissioner, it will not be possible for the commissioner to withdraw the document. The charge may only be disposed of by trial or by action of the State's Attorney.

You will be required to appear at the trial as a witness. Failure to appear on the date set by the court could result in your arrest for failure to obey a court order.

The application which you are filing is being filed under oath. Criminal Law Article § 9-503, of the Annotated Code of Maryland provides that any person who makes a false statement or report of a crime or causes such a false report or statement to be made to any official or agency of this State, knowing the same, or any material part thereof, to be false and with intent that such official or agency investigate, consider or take action in connection with such statement or report, shall be subject to a fine of not more than $500, or be imprisoned not more than six months, or be both fined and imprisoned, in the discretion of the court.

It is essential that your furnish as much information as possible about the offense. To be sure that your information is adequate, your application should clearly state the following:

1.  WHO?
    Identify the accused, (the person you are complaining about), and identify yourself.

2.  WHEN?
    The time, day, month and year of the offense.

3.  WHERE?
    The exact address and street, the city, county and state where the offense happened. Also state whether the offense happened in a private home or in some public place.

4.  WHAT?
    State exactly what was done to you. For example: if property was taken, describe it and its value; or, if property was damaged or destroyed, indicate the original cost of the item or its replacement value. If you do not know the exact value, estimate it as accurately as possible.

5.  WHY?
    The facts you give must show the accused intended to commit a criminal act.

6.  HOW?
    How the accused committed the offense. For example, if you were physically assaulted, were you struck with a fist, a flat hand, kicked, or pushed, or were you struck with an object, such as a club or pipe, etc.? If property was taken, how did the accused get it? If it was destroyed or damaged, how did the accused cause the damage?

7.  At the top of the application, you will notice a space marked "DESCRIPTION". The information in this space refers to the **accused**. It is important that you furnish as much of this as possible so that the accused may be easily identified.

If you need further assistance in completing your application, please feel free to ask the commissioner.

You are entitled to request that address and telephone number of a victim, a complainant, or a witness be considered for shielding at the filing of this application.

NOTICE: Remote access to the name, address, telephone number, date of birth, e-mail address, and place of employment of a victim or non-party witness is blocked. (Md Rule 16-1008(a)(3)(B))

# Petition for Expungement of Record

MARYLAND
JUDICIARY

☐ **CIRCUIT COURT** ☐ **DISTRICT COURT OF MARYLAND FOR** _____
<br>City/County

Located at _____ Case No _____
<br>Court Address

Tracking # _____

STATE OF MARYLAND                    vs. _____ ___ ___
<br>Defendant                                    DOB

## FORM 4-504.1. PETITION FOR EXPUNGEMENT OF RECORDS

1. (Check one of the following boxes) On or about _____, I was ☐ arrested, ☐ served with a summons,
<br>Date

or ☐ served with a citation by an officer of the _____
<br>Law Enforcement Agency

at _____, Maryland, as a result of the following incident _____

_____

2. I was charged with the offense(s) of _____

3. On or about _____, the charge was disposed of as follows (check one of the following boxes):
<br>Date

☐ I was acquitted and either three years have passed since disposition or a General Waiver and Release is attached.

☐ The charge was dismissed or quashed and either three years have passed since disposition or a General Waiver and Release is attached.

☐ A judgment of probation before judgment was entered on a charge that is not a violation of Code*, Transportation Article, § 21-902 or Code*, Criminal Law Article, §§ 2-503, 2-504, 2-505, or 2-506, or former Code*, Article 27, § 388A or § 388B, and either (a) at least three years have passed since the disposition, or (b) I have been discharged from probation, whichever is later. Since the date of disposition, I have not been convicted of any crime, other than violations of vehicle or traffic laws, ordinances, or regulations not carrying a possible sentence of imprisonment; and I am not now a defendant in any pending criminal action other than for violation of vehicle or traffic laws, ordinances, or regulations not carrying a possible sentence of imprisonment.

☐ A Nolle Prosequi was entered and either three years have passed since disposition or a General Waiver and Release is attached. Since the date of disposition, I have not been convicted of any crime, other than violations of vehicle laws, ordinances, or regulations not carrying a possible sentence of imprisonment; and I am not now a defendant in any pending criminal action other than for violation of vehicle or traffic laws, ordinances, or regulations not carrying a possible sentence of imprisonment.

☐ The proceeding was stetted and three years have passed since disposition. Since the date of disposition, I have not been convicted of any crime, other than violations of vehicle or traffic laws, ordinances, or regulations not carrying a possible sentence of imprisonment; and I am not now a defendant in any pending criminal action other than for violation of vehicle or traffic laws, ordinances, or regulations not carrying a possible sentence of imprisonment.

☐ The case was compromised or dismissed pursuant to Code*, Criminal Law Article, § 3-207, former Code*, Article 27, § 12A-5, or former Code*, Article 10, § 37 and three years have passed since disposition.

☐ On or about _____, I was granted a full and unconditional pardon by the Governor for the one
<br>Date
criminal act, not a crime of violence as defined in Code*, Criminal Law Article, § 14-101(a), of which I was convicted. Not more than ten years have passed since the Governor signed the pardon, and since the date the Governor signed the pardon I have not been convicted of any crime, other than violations of vehicle or traffic laws, ordinances, or regulations not carrying a possible sentence of imprisonment; and I am not now a defendant in any pending criminal action other than for violation of vehicle or traffic laws, ordinances, or regulations not carrying a possible sentence of imprisonment.

WHEREFORE, I request the Court to enter an Order for Expungement of all police and court records pertaining to the above arrest, detention, confinement, and charges.

I solemnly affirm under the penalties of perjury that the contents of this Petition are true to the best of my knowledge, information and belief, and that the charge to which this Petition relates was not made for any nonincarcerable violation of the Vehicle Laws of the State of Maryland, or any traffic law, ordinance, or regulation, nor it is part of a unit the expungement of which is precluded under Code*, Criminal Procedure Article, § 10-107.

*References to "Code" in this Petition are to the Annotated Code of Maryland.

_____  _____        _____  _____
Signature of Attorney       Date             Signature of Defendant      Date

_____                     _____
Name - Printed                               Name - Printed

_____                     _____
Address                                      Address

_____  _____   _____  _____
Telephone Number                             Telephone Number

**CC-DC/CR 72** (Rev. 8/2006)        | Reset |

☐ **CIRCUIT COURT** ☐ **DISTRICT COURT OF MARYLAND FOR** _____
City/County

Located at _____  Case No. _____
Court Address

STATE OF MARYLAND  vs.  _____
Defendant

_____
Address

_____
City, State, Zip

_____  _____
Telephone No. - Home  Telephone No. - Work

## GENERAL WAIVER AND RELEASE

I, _____ , hereby release and forever

discharge _____ , and the _____ ,
Complainant  Law Enforcement Agency

all of its officers, agents, and employees, and any and all other persons from any and all claims which I may

have for wrongful conduct by reason of my arrest, detention, or confinement on or about _____ .
Date

This General Waiver and Release is conditioned on the expungement of the record of my arrest,

detention, or confinement and compliance with Code*, Criminal Article, § 10-103(c) or § 10-105, as

applicable, and shall be void is these conditions are not met.

WITNESS my hand and seal this _____ day of _____ , _____ .
Month  Year

TESTE:

_____  _____(Seal)
Witness  Signature

\* References to "Code" in this Petition are to the Annotated Code of Maryland.

# CLASSROOM ASSIGNMENTS & STUDENT STUDY PROJECTS

## CLASSROOM ASSIGNMENTS & STUDENT PROJECTS

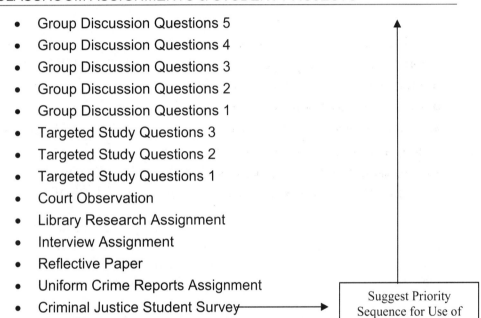

- Group Discussion Questions 5
- Group Discussion Questions 4
- Group Discussion Questions 3
- Group Discussion Questions 2
- Group Discussion Questions 1
- Targeted Study Questions 3
- Targeted Study Questions 2
- Targeted Study Questions 1
- Court Observation
- Library Research Assignment
- Interview Assignment
- Reflective Paper
- Uniform Crime Reports Assignment
- Criminal Justice Student Survey
- Sample Course Description & Learning Objectiv

Suggest Priority Sequence for Use of Assignments

**Note:** Course Assignments are organized in the suggested priority sequence for use from the back of the supplement forward to make it convenient for students to remove assignments with limited impact on the actual Supplement binding.

**CRIMINAL JUSTICE**
**GROUP DISCUSSION QUESTIONS**
**DISCUSSION QUESTION NO. 5 [___Points]**

**Subject:** *Bail Review*

Please review the following fact scenario, answer the questions presented, and be prepared to discuss the facts and your responses with your classmates.

**Instructions:**
Review the following fact scenario and formulate your answers. Discuss the questions with your classmates if this assignment is part of a group exercise, and compare the group's responses with your own answers.

**Facts:**
The City of Capital has experienced an unprecedented number of drug offenses since 2007. On a sunny day, three young men from New York ages 17, 19 and 24 arrive at the train station and are immediately stopped by the police. Eventually the police search a gym bag being carried by on of the young men, and find 1 kilo of marijuana inside.

All three young men are arrested and charged with possession with intent to distribute marijuana. At the bail review, bail for the 17 yr old, Justin is set at $500,000.00 dollars, although Justin has no criminal record at all. However, the judge indicated that he wanted to send a strong message to young drug dealers who visit the City of Capital.

Bail for the second defendant, Harvey, age 23 was set at $25,000.00 dollars. Harvey is a college student with minor traffic violations. Bail for the third suspect, Oscar, age 35 was set at no bail.

**Questions:**
1. Are there any constitutional issues raised by the judge's imposition of different bail amounts for Justin, Harvey and Oscar? Discuss each bail issue and support your answers with solid reasons.

2. Is the judge's method of sending a message to young men through the use of high bail, and no bail, a fair or appropriate exercise of the juridical power? Explain your answer.

3. Provide your responses in a typewritten paper and turn them into the instructor.

**CRIMINAL JUSTICE**
**GROUP DISCUSSION QUESTIONS**
**DISCUSSION QUESTION NO. 4 [___Points]**

**Subject:** *Private Security Officer - Search & Seizure*

Please review the following fact scenario, answer the questions presented, and be prepared to discuss the facts and your responses with your classmates.

**Instructions:**
Review the following fact scenario and formulate your answers. Discuss the questions with your classmates if this assignment is part of a group exercise, and compare the group's responses with your own answers.

**Facts:**
Jack, a private security officer, became increasingly concerned about the large number of visitors arriving at and leaving the townhouse of his next door neighbor Rob, on Roughshod Lane in Denton, Maryland. After a month of observing the increasing foot traffic to Rob's house, Jack suspected the his neighbor was dealing drugs. One day Jack decided to do something about the situation and climbed through an open window at the rear of Rob's townhouse. He looked inside while Rob was gone to the store and found 10 grams of cocaine in Rob's bedroom. Jack immediately called the local police. The police arrived and Rob turned the cocaine over to them. Upon Rob's return to the townhouse, he was arrested by the police for possession of with intent to distribute cocaine.

**Questions:**
1. Can the cocaine discovered by Jack be introduced into evidence by the police at trial?

2. Was Jack's search illegal? Why or why not?

3. What lawsuit causes of action does Rob have against Jack?

4. Provide your responses in a typewritten paper and turn them into the instructor.

**CRIMINAL JUSTICE**
**GROUP DISCUSSION QUESTIONS**
**DISCUSSION QUESTION NO. 3** [___Points]

**Subject:** *Police Gratuities*

Please review the following fact scenario, answer the questions presented, and be prepared to discuss the facts and your responses with your classmates.

**Instructions:**
Review the following fact scenario and formulate your answers. Discuss the questions with your classmates if this assignment is part of a group exercise, and compare the group's responses with your own answers.

**Facts:**

Jerry is a 7 year police patrol officer who likes his job in the Maryland suburbs. Over the years, people in the community and business owners have gotten to know and like Jerry and routinely invite him to their convenience stores, fast food restaurants, and the local diner to eat or have coffee, which helps to make them feel safe when he is taking his on duty lunch break.

Typically, local merchants will give Jerry a police officer's discount of 50% on meals and free coffee for the night shift work each week that Jerry is on duty. Jerry works 5 days a week, 8 hours a day and takes a lunch break and a coffee break on each shift.

Jerry expects to retire within 13 years.

**Questions:**
1. Is it okay for Jerry to receive free coffee? Give reasons for your answer.

2. Is it okay for Jerry to receive half price meals as part of a police officer's discount? Give specific reasons for your answer.

3. Does Jerry's acceptance of coffee or half price meals compromise his ability to carry out his official responsibilities in any way? Discuss the issues that may arise under these circumstances.

4. Provide your responses in a typewritten paper and turn them into the instructor.

**CRIMINAL JUSTICE**
**GROUP DISCUSSION QUESTIONS**
**DISCUSSION QUESTION NO. 2 [___Points]**

**Subject:** *Patriot Act Legislation*

Please review the following fact scenario, answer the questions presented, and be prepared to discuss the facts and your responses with your classmates.

**Instructions:**
Review the following fact scenario and formulate your answers. Discuss the questions with your classmates if this assignment is part of a group exercise, and compare the group's responses with your own answers.

**Facts:**

1. Review your text concerning *Procedural Law* which is designed to safeguard the accused from potential abuses by the Government and instill fairness and due process into the criminal judicial process.

2. Look in your text and review the Patriot Act provisions. Go online to **google,** review and compare the current controversies surrounding the latest version of the Patriot Act legislation as it relates to (i.e. police secret home searches, interception of emails, obtaining library patron reading records, and taking other actions - all without notice or necessity of obtaining a judicial warrant, and/or preventing disclosure to the person being investigated)

3. Formulate your own thoughts and concerns, if any, that you may have concerning the government's authority or use of authority under the current Patriot Act legislation.

4. In light of your reading and review, discuss how we can protect the accused from governmental abuse of the law in accordance with the U.S. Constitution (i.e. 4th Amendment Search & Seizure Protection, 6th Amendment Right to Counsel & Right To Confront Your Accuser, etc.)

5. Also provide a recommendation as to how we can protect our cherished constitutional rights while balancing the safety needs of the government and the rights of persons who have been victimized by terrorist criminal activity.

6. Provide your responses in a typewritten paper and turn them into the instructor.

## CRIMINAL JUSTICE
## GROUP DISCUSSION QUESTIONS
## DISCUSSION QUESTION NO. 1 [___Points]

**Subject:** *Barefoot Driving*

Please review the following fact scenario, answer the questions presented, and be prepared to discuss the facts and your responses with your classmates.

**Facts:**
Cameron is driving with his two friends en route to Virginia Beach for a final summer vacation before the school semester starts. A State Trooper pulls Cameron's car over on Interstate 95 (I-95) just south of the Wilson Bridge to verbally warn Cameron about the loose bike rack on his car. The trooper approaches the drivers' side of the vehicle and notices that Cameron is driving barefoot?

**Instructions:**
Review the following questions below and formulate your answers. Discuss the questions with your classmates if this assignment is part of a group exercise, and compare the group's responses with your own answers.

**Questions:**

1. Did Cameron violate the Maryland law by driving barefoot?
   a. If so, identify the applicable law and provide reasons and a reference source for your answer.
   b. If not, give reasons for your answer and provide a reference source for your answer.

2. Can the Trooper validly issue a ticket to Cameron?

3. Did you know the answer to this question before taking this class? If so, state the source of your information prior to discussing this scenario in class.

4. If Cameron gets a traffic violation ticket, how should the Court rule of her Not Guilty defense? Guilty or Not Guilty?

5. Explain any other conclusions you have reached that you believe support your answer.

Provide your responses in a typewritten paper and turn them into the instructor.

**CRIMINAL JUSTICE**
**TARGETED COURSE COMPREHENSION**
**STUDY QUESTIONS 3**
[___Points]

**Chapters 8-17 or _____-_____**

**Briefly explain or answer each question below.** These **Study Questions 3** are designed to assist you in developing and in-depth understanding of the Criminal Justice subjects covered during the last one third of your course text, and should be used as a supplemental preparation tool to help familiarize you with the course content that may appear on your course examination.

In a few instances, there may be a few overlap questions for topics covered in study questions 1 or 2. Your instructor will determine whether and to what extend the assignment may be counted for credit according to the course grading scale. Your instructor will also determine whether and to what extent the completed study questions may be considered as part of your participation credit in the course learning process. See the Course Schedule for Due date.

1. Define what is meant by the term "indigent person"?

2. What name is commonly used to refer to large city prosecutors?

3. Identify and explain three common types of pleas that defendants can enter at arraignment?
   a.

   b.

   c.

4. How many jurors do criminal defendants get at trial?

5. How much of a criminal sentence must a Maryland inmate serve before being eligible for parole?

1 of 9

6. What is determinate sentencing?

7. What are the criticisms of determinate sentencing?

8. What is indeterminate sentencing?

9. What are the criticisms of indeterminate sentencing?

10. Under what circumstances must police officers obtain an arrest warrant before making an arrest?

11. Explain the process for getting appointed as a judge?

12. Explain what rights victims have to participate in the criminal justice process.

13. What are sentencing guidelines?

    a. Explain the characteristics of federal sentencing guidelines since 2007? (mandatory or discretionary)

    b. Explain the characteristics of state sentencing guidelines? (mandatory or discretionary)

14. Describe the criteria used by prosecutors in determining whether to charge a person with a criminal offense?

15. What is an indictment?

16. Explain the costs of capital punishment cases (death penalty) as compared with the costs of keeping a convicted inmate in prison for life without parole?

17. What is plea bargaining?


18. What percentages of cases are resolved by plea bargaining before trial in the criminal justice system?


19. What is a Writ of Certiorari and explain the percentage of Writ of Certiorari cases traditionally denied by the U.S. Supreme Court?


20. What is bail?


21. Explain the typical percentage of bail money a person must pay a bail bondsman when posting bail in a criminal case?


22. Is the money that was posted for bail in a criminal case returned after the accused voluntarily shows up in court on the date of trial? Explain.


23. Explain the following punishment terms:

    a. Deterrence-


    b. Incapacitation-


    c. Rehabilitation-


    d. Retribution-


    e. Revenge-


24. What is allocution in a criminal case?


25. What is mitigation in a criminal case?

26. What is a court of general jurisdiction?

27. What is a court of limited jurisdiction?

28. List and Explain the critical stages of the criminal justice process?

29. What is a public defender?

30. What is a court appointed lawyer?

31. Name at least two types of private defense attorneys.

32. Where do the courts get the pool of potential persons to serve as jurors in criminal cases?

33. Explain who is the least likely group of persons in society to serve on the jury panel?

34. What is a writ of habeas corpus?

35. What is an "arraignment"?

35. Explain the basic goal(s) of early punishment for crimes and give two examples

36. Define and distinguish the following different forms of punishment :
   a. Banishment -

   b. Transportation -

   c. Workhouses -

37. Describe the characteristics of the panopticon prison design.

38. Explain the difference between the Pennsylvania system and the Auburn system type penitentiaries.

39. What is the medical model of correction?

40. Explain the concept and purpose of privatization of correctional facilities.

41. Explain shock incarceration and the characteristics of boot camp facilities.

42. What is a chain gang?

43. How do we determine the incarceration rate?

44. What is a classification facility?

45. Define security level.

46. Define custody level.

47. What is a "Big House" correctional facility?

48. Under the Reformatory Movement what is one of the principle advantages associated with an indeterminate sentence according the text?

49. What is a lockup?

50. Define the meaning of jail and John Irwins' findings concerning the "rabble class".

51. Define new generation jail.

52. Explain the difference between the following:
    a. protective custody -

    b. administrative segregation -

53. What are conjugal visits?

5 of 9

54. Explain the snitch system and how it functions.

55. What does the term "Total Institution" mean?

56. Explain the Convict Code.

57. Explain the Deprivation Model.

58. Define Prisonization

59. Explain the Importation Model.

60. Explain some of the reasons given for the high rates of prison violence.

61. What is the "Sub-rosa" economy.

62. What types of vulnerable inmates exist in prison societies?

63. Explain the types of prison victimization that occurs amongst inmates.

64. Define and distinguish the inmate coping styles of:
    a. Doing Time-

    b. Jailing-

    c. Gleaning-

65. What are Pseudofamilies?

66. Explain some of the conflicts faced by correctional officers in their work.

67. Explain the "hands off " philosophy of the courts concerning inmate rights.

68. Define Habeas Corpus.

69. What is a jailhouse lawyer?

70. What procedural rights do inmates have when faced with the loss of "good time credits"

71. What free speech restrictions can prison officials place upon inmates?

72. What religious restrictions can prison officials place upon inmates?

73. What "constitutional standard" must an inmate meet under the 8th Amendment to be successful in a case involving the alleged lack of medical care by prison officials?

74. Explain the meaning of the phrase "Total Prison Conditions".

75. What is a commutation?

76. What is parole?

77. What is good time?

78. Define the meaning of mandatory release (dates).

79. What is recidivism?

80. Define Community Corrections.

81. Why is community corrections important today?

82. Define probation and how it developed?

83. Explain the distinctions between the following types of probation:
    a. straight probation-
    b. suspended –sentence probation -
    c. split sentence-
    d. shock probation-
    e. residential probation-

84. Explain the purpose of a Presentence Investigation Report (PSI) and how does it correspond with the actual sentence imposed by the court?

85. What are probation conditions?

86. Explain how a parole release status can be terminated:
    a. Revocation -
    b. Technical violations -

87. Explain why probation fees may be included as part of a defendants' probation sentence?

88. What is recidivism?

89. What is the purpose of the parole board?

90. Explain reintegration?

91. Define a day reporting center.

92. What is home confinement?

93. What is electronic monitoring?

94. Explain the purpose of a halfway house?

95. What is a temporary release program?

96. Define juvenile delinquency.

97. Explain the apprenticeship system.

98. Explain the binding-out system.

99. Define "House of Refuge"

100.    What is placing out?

101.    What is a reform school?

102.    Explain cottage reformatories.

103.    Why did the Juvenile Court system develop during the Progressive Era?

104.    Define the theory of "Parens Patriae".

105.    How were early Juvenile Hearings conducted under the Juvenile Court Act of 1899 in Chicago.

106.         Explain the significance of the In re Gault Decision?

107.         What is adjudication?

108.         Describe Informal Juvenile justice.

109.         What is a status offense?

110.         What is the purpose of intake screening in juvenile cases?

111.         Explain the process of juvenile [transfer, waiver or certification] in certain juvenile proceedings.

112.         Explain the two types of juvenile hearings.

113.         What is a "disposition" in juvenile court?

114.         Explain some of the problems associated with juvenile detention facilities.

115.         Define restitution?

END OF STUDY QUESTIONS 3

9 of 9

**CRIMINAL JUSTICE**
TARGETED COURSE COMPREHENSION
STUDY QUESTIONS 2
[___Points]

**Chapters 4-8** or _____ - _____

**Briefly explain or answer each question below.** These **Study Questions 2** are designed to assist you in developing and in-depth understanding of the Criminal Justice subjects covered during the second one third of your course text, and should be used as a supplemental preparation tool to help familiarize you with the course content that may appear on your course examination.

In a few instances, there may be a few overlap questions for topics covered in study questions 1 or 3. Your instructor will determine whether and to what extend the assignment may be counted for credit according to the course grading scale. Your instructor will also determine whether and to what extent the completed study questions may be considered as part of your participation credit in the course learning process. See the Course Schedule for Due date.

1. What is mens rea?

2. What is actus reus?

3. Give the definition of a crime?

4. Identify three indexes used to report and survey the occurrences of crime.

5. How do we calculate the crime rate?

6. Explain the characteristics of the Crime Control Model of Criminal Justice.

1 of 5

7. Explain the characteristics of the Due Process Model of Criminal Justice.

8. Identify and explain who is considered to be the father of modern American policing? Give your supporting reason(s).

9. What is role conflict?

10. Describe the following types of patrol:
    Aggressive-

    Directive –

    Full enforcement-

    Selective enforcement-

11. Explain the theory of "broken windows" by James Q. Wilson.

12. What is typical the name of the state court that receives death penalty appeals after trial of persons convicted and sentenced to death?

13. Explain the three functions of police work?

14. Explain how police authority may be limited?

15. What did August Vollmer advocate?

16. Define Meat Eaters?

17. Define Grass Eaters?

18. Define police corruption.

19. What is a field interrogation or field interview?

20. Define community policing and explain how it works?

21. What is a charging document? Give three examples?
    a.

    b.

    c.

22. Explain why police departments now prefer college educated recruits?

23. What is police "role expectation"?

24. Explain the three police operational styles described by James Q. Wilson?
    a.

    b.

    c.

25. Explain some of the possible requirements used for new police recruit selection?

26. Identify and describe the factors that affect the use of police discretion?

27. What is racial profiling & how does it affect the use of police discretion?

28. Define excessive police force?

29. Explain when police can use deadly force?

30. Briefly describe some of the new powers given to law enforcement agencies under the USA Patriot Act?

   a.

   b.

   c.

   d.

   e.

31. Identify and briefly describe the several agencies centralized under the Department of Homeland Security U.S.?

   1.

   2.

   3.

   4.

   5.

   6.

32. What is the FBI definition of terrorism?


33. According to your text, what is the definition of terrorism under the U.S. Code of Federal Regulations 28 C.F.R. Section 0.85?


34. What is the definition of terrorism under Title 22 of the U.S. Code Section 2656f (d)?


35. Identify the (4) key elements of terrorism?

    1.

    2.

    3.

    4.


END OF STUDY QUESTIONS 2

## CRIMINAL JUSTICE
## TARGETED COURSE COMPREHENSION
## STUDY QUESTIONS 1
[___Points]

**Chapters 1-5 or _____ - _____**

**Briefly explain or answer each question below.** These **Study Questions 1** are designed to assist you in developing and in-depth understanding of the Criminal Justice subjects covered during the first one third of your course text, and should be used as a supplemental preparation tool to help familiarize you with the course content that may appear on your course examination.

In a few instances, there may be a few overlap questions for topics covered in study questions 2 or 3. Your instructor will determine whether and to what extend the assignment may be counted for credit according to the course grading scale. Your instructor will also determine whether and to what extent the completed study questions may be considered as part of your participation credit in the course learning process. See the Course Schedule for Due date.

1. Explain the effect the media has on the reporting of crime.

2. What is the purpose of the criminal justice system?

3. Identify and explain three institutions of social control

4. Define law.

5. Explain the use of law as a means of formal social control.

6. Define and explain court jurisdiction.

7. Define and explain criminal justice agency's jurisdiction.

8. What is the Grand Jury?

9. What is a victimless crime?

1 of 7

10. Identify three types of charging documents and describe how each one is used.

11. Define a misdemeanor.

12. Define a felony.

13. What is the purpose of Miranda Warnings and When must they be given to a criminal suspect?

14. Explain the purpose of a preliminary hearing and when does it apply?

15. Explain how a bench trial or summary trial works.

16. Define probable cause.

17. Explain what constitutes a detention or investigatory stop.

18. What is a plea bargain and what percentage of criminal cases end in a plea bargain prior to trial?

19. Who is Sir Robert Peel? And what were his accomplishments?

20. Explain the distinction between parole and probation.

21. What is the definition of a crime?

22. What is a norm?

23. Define mens rea.

24. Define actus reus.

25. Define Mala In Se Crimes.

26. Define Mala Prohibita Crimes.

27. What is an ex post facto law?

28. Identify and explain the defenses of infancy, necessity, entrapment and self defense.

29. Explain the insanity defenses available to a criminal defendant
    a. **McNaughten-**

    b. **Substantial Capacity-**

    c. **Irrestible Impulse-**

    d. **Maryland – NCR Plea (Not Criminally Responsible) -**

30. Define status offense?

31. What is the (UCR)-Uniform Crime Report ? Explain the disadvantages?

32. What is the dark figure of crime?

33. What is the (NCVS)-National Crime Victimization Survey ? Explain the advantages and disadvantages?

34. Define status offense?

35. Define juvenile delinquency .

36. Explain the positivist theory of crime.

37. Explain the classical theory of crime.

38. Explain the criminological theory.

39. Explain the social contract.

40. Who is Cesare Becarria ? What is his theory of crime causation and punishment?

41. Who is Cesare Lombroso ? What is his biological theory of crime causation and punishment?

42. Explain the concept of Utility.

43. Define General Deterrence.

44. Explain the underlying basis of the psychological theories of crime causation.

45. Explain Sigmund Freud's psychoanalytic theory of crime

46. Define sociopath & psychopath.

47. Identify and Explain Maslow's Hiearchy of Needs.

48. What is Anomie ?

49. Explain Emile Durkheim's sociological theory of crime causation.

50. Define Social Disorganization.

51. Explain Strain Theory.

52. Explain Learning Theory.

53. Explain Differential Association Theory.

54. Explain Control Theory.

55. Explain Conflict Theory.

56. Explain Labeling Theory.

57. Explain Class Struggle Theory.

58. Explain Feminist Theory.

59. Define Criminal Law.

60. What is the Penal Code?

61. What is a Tort?

62. Define Civil Law.

63. What is substantive Law?

64. What is procedural Law?

65. Identify and **list at least** (5-7) characteristics of criminal law.

66. Define Common law and briefly explain the origins of its development.

67. Explain the U.S. Constitution and Legislative Statutes as sources of law

68. Define Precedent and Stare Decisis.

69. What is the 14$^{th}$ Amendment & what protections does it provide?

70. Identify the **Bill of Rights** and explain how they eventually became applicable to the states?

71. What is the definition of a search?

72. What is a seizure?

73. What is a Warrant?

74. Define Arrest.

75. Explain whether warrantless searches of a home are legal or illegal?

76. Define Contraband.

77. What is a search incident to an arrest?

78. What area can a police officer search under the case of **Chimel v California?**

79. What is Mere Suspicion?

80. Define Reasonable or Articulable Suspicion?

81. What is a Frisk or Pat Down ? and When may police officers conduct a frisk or pat down?

82. Explain the differences between the "beyond a reasonable doubt" criminal case standard & the "preponderence of the evidence" civil case standard of proof.

83. What is the exclusionary rule and how do the courts apply the rule to wrongful police conduct?

84. What is the good-faith exception to the exclusionary rule and how do the courts apply the rule to police conduct?

85. Explain Double Jeopardy and Identify the Constitutional Amendment that protects individuals from Double Jeopardy circumstances.

86. Explain what is meant by Self Incrimination?

# CRIMINAL JUSTICE
## LIBRARY RESEARCH ASSIGNMENT
## LIST OF CASES

1. The Maryland Criminal Cases for this assignment are sorted below by name and case citation number which can be used to locate the case using the Lexis Nexis research database.

2. <u>Select One</u> case from the following list and record the case name and case number on Page 1 of the library research assignment. Turn in the name of the case that you have selected to the instructor, or send it to the instructor by email at

  _____.

4. Complete the written library research assignment by answering the questions contained on Page 1 of the Library Research Assignment Sheet:

### <u>LIST OF MARYLAND CRIMINAL CASES</u>

| CASE NAME | CASE CITATION |
| --- | --- |
| **2007 CASES** | |
| SPRY V. STATE | 396 MD. 682 |
| FIELDS V. STATE | 172 MD.APP. 496 |
| CHRISTIAN V. STATE | 172 MD.APP. 212 |
| ROBEY V. STATE | 397 MD. 449 |
| CHANEY V. STATE | 397 MD. 460 |

1 of 3

| HARRIS V. STATE | 173 MD. APP. 71 |
| MASSEY V. STATE | 173 MD. APP. 94 |
| WILLIAMS V. STATE | 173 MD. APP. 161 |

| 2006 CASES | CASE CITATION |
| --- | --- |
| SINGFIELD V. STATE | 172 MD. APP. 168 |
| ROLLINS V. STATE | 172 MD. APP. 56 |
| ALLEN V. STATE | 171 MD. APP. |
| TWINE V. STATE | 395 MD. 539 |
| MYERS V. STATE | 395 MD. 261 |
| RUFFIN V. STATE | 394 MD. 355 |
| RIVERS V. STATE | 393 MD. 569 |
| TOTH V. STATE | 393 MD. 318 |
| KANG V. STATE | 393 MD. 97 |
| GRINER V. STATE | 168 MD. APP. 714 |
| KELLY V. STATE | 392 MD. 511 |
| SURLAND V. STATE | 392 MD. 17 |

| | |
|---|---|
| PEREZ V. STATE | 168 MD.APP. 248 |
| THOMPSON V. STATE | 167 MD.APP. 513 |
| WALKER V. STATE | 391 MD. 233 |
| STATE V. PITT | 390 MD. 697 |
| BROWN V. STATE | 168 MD.APP. 400 |
| BRYANT V. STATE | 393 MD. 196 |
| CHOW V. STATE | 393 MD. ⌐1 |
| SMITH V. STATE | 3⌐⌐ MD. 184 |
| WILLIAMS V. STATE | 394 MD. 98 |
| GARRETT V. STATE | 394 MD. 217 |
| OWENS V. STATE | 170 MD.APP. 35 |
| ⌐RE V. STATE | 170 MD.APP. 1 |

3 of 3

149

_____ [College]
**Introduction to Criminal Justice**
_____ [Instructor]

REFLECTIVE PAPER
INTERNET –BASED WRITING ASSIGNMENT

The purpose of this Reflective Paper is to focus your attention on a significant problem or challenge existing in the administration of Criminal Justice today.

You will be graded on the overall qualitative content of your Paper, your analysis of the identified problem and the quality of your proposed solution or resolution of the problem.

Proper grammar, spelling and sentence structure is expected on all papers.

The Original Reflective Paper should be a minimum of (4-5) pages in length, typewritten and should have a Title Page with your Name, Topic, Class and Meeting Time. Sources should be footnoted and included in a Works Cited page at the end of your paper.

In the Original Reflective Paper you will be required to:

1. Identify a significant problem area that exists in Criminal Justice (i.e. courts, crimes, juvenile justice system, probation, law enforcement, sentencing, discrimination, minority officers, etc. ) based upon your readings, class discussions, your life experiences, or the life experiences of someone that you know.

2. Describe in detail, the nature and extent of the problem in society.

3. Read and briefly summarize in your own words, 2-4 reference sources (using ONLY the Lexis/Nexis Academic Internet Database or the Local Newspaper\ Journal Database Sources available at the college) that support, oppose or discuss your
identified problem.

a. Properly identify and credit the sources in your paper and attach a copy of the references to your completed paper.

4. Analyze and discuss how this problem in any way affects your life or the life of someone that you know. Give clear details and examples.

5. Provide a proposed solution or resolution of the problem that you have identified.

See Course Schedule or instructor for due date. _____

**CRIMINAL JUSTICE**
**UNIFORM CRIME REPORTS ASSIGNMENT** [___Points]

**Instructions:**

Research the following crime statistics using the Maryland Uniform Crime Reports. Go online to http://www.goccp.org/four/research/ucr/ucr.php or to:

_____to complete this assignment.

Research the following:

## I.  2006 UCR Uniform Crime Reports - Maryland

1. In 2006 what were the total number of Violent Crimes "actually reported" in Maryland under the index of Crime-State 2006?_____.

2. What were the total number of Motor Vehicle thefts in Maryland under the Index of Crime-State **2006**_____

3. List the total number of **2006** "Rapes" in Maryland, as reported by the following Counties:

a. Howard County Police_____
b. Montgomery County Police_____
c. Baltimore County Police_____
d. Anne Arundel Police Dept._____
e. Prince George's Co. Police_____
f. Carroll Co. State Police_____
g. Queen Anne's County _____
h. Talbot County _____
i. Wicomico County _____
j. Dorchester _____
k. Charles County _____
l. Frederick County _____
m. Alleghany County _____
n. Cecil County _____

4.    In 2006- List the number of "Breaking & Entering" incidents in Maryland in the following counties as reported by:
a.    Howard County Police_____
b.    Montgomery County Police_____
c.    Baltimore County Police_____
d.    Anne Arundel Police Dept._____
e.    Prince George's Co. Police_____
f.    Carroll Co. State Police_____
g.    Queen Anne's County _____
h.    Talbot County _____
i.    Wicomico County _____
j.    Dorchester _____
k.    Charles County _____
l.    Frederick County _____
m.    Alleghany County _____
n.    Cecil County _____

5.    **In 2006** – List the name of your Maryland county here [_____]
and list the number of persons "Arrested" in your county in Maryland in each of the following categories:

o.    Driving Under Influence (DUI)          _____

p.    Drug Possession (CDS)          _____

q.    Weapons Violations          _____

r.    Domestic Violence          _____

s.    Sex Offenses          _____

| College | Instructor | Student |
|---|---|---|

**CRIMINAL JUSTICE STUDENT SURVEY**

1. Name_____

2. address _____

3. Home Phone

4. Work Phone

5. Cell Phone _____

5a. Email Address _____

6. Your college major

7. Current work or position_____

5. Your outside interests

6. Why you are taking this class?

7. What you expect to get out of this class?

8. What you can contribute to learning in the course?

9. Number of total college credits completed _____

10. Number of college credits completed at _____ (college)

11. approximate age ranges

　　　　18-24 _____
　　　　25-35 _____
　　　　45-55 _____
　　　　above 55 _____

*Optional: Age data is used only for statistical course evaluation purposes or outcomes assessment course evaluation purposes

12. Will this course help you in you future career?

Explain:_____

13. Will this course help you in your daily life?


Explain:_____

**Check One:**

14. ____ Day Student      ____ Evening Student      _____ Online

15. List other courses that you are interested in taking at _____ college?

_____

_____

_____

_____

_____

_____

## COURSE DESCRIPTION AND OUTLINE

# _____ 101

# Introduction to Criminal Justice

### 3 Semester Hours

_____ [Name]
_____ [Professor/Adjunct]
_____

Office: _____
Phone: (    ) _____
       (voicemail)
Email: _____

Office hours
Monday - _____
Tuesday - _____
Wednesday - _____
Thursday- _____
Friday
_____**[or by appointment]**

## Course Description

Introduction to Criminal Justice is survey of the history, philosophy and social development of police, courts and corrections in a democratic society. Identification and operations of local, state and federal agencies will be covered with a criminal justice career orientation.

**The text is:**_____, Introduction to Criminal Justice (____Ed.)
      [Author]

_____. (_____).
      [Publisher]

This course will require additional outside assignments and readings not included in the text, and will require students to visit a court room criminal trial, interview a criminal justice professional, and conduct at least one library research project at. Each individual assignment will be provided with separate instructions. Check the due dates for each assignment frequently. Students will be responsible for working independently, reviewing all assigned materials and should not rely solely upon lectures or classroom discussions to provide them with a complete understanding of the course materials.

## Objectives

Upon successfully completing this course, the student should be able to:

1. Define crime and understand its elements according to legal definition.
2. Describe the administration of justice through stages of investigation, arrest, booking, initial appearance, preliminary hearing, indictment, arraignment, trial, and sentencing.
3. Examine crime statistics through the use of UCR, NCS and self report surveys and specifically analyze the crime statistics of Maryland and your County.
4. Identify methods of research and the limitations of each method.
5. Examine crime control versus due process both theoretically and as it relates to public policy.
6. Describe the major crime causation theories and the public policy responses to each.
7. Distinguish between substantive and procedural criminal law and elements of each.
8. Identify the criminal defenses and their relation to mens rea.
9. Identify the $4^{th}$, $5^{th}$, $6^{th}$, $8^{th}$, and $14^{th}$ amendments and major court cases related to each.
10. Analyze "search and seizure", exclusionary ruling, and warrantless searches.
11. Examine the history and functions of police.
12. Examine and evaluate policing issues and trends.
13. Examine the adult trial and post-trial process, including the role of the prosecutor and defense attorney.
14. Distinguish between state and federal court jurisdiction and venue.
15. Examine the history and functions of corrections.
16. Examine the juvenile court proceedings, the trial process, and the post-trial procedures.
17. Examine and evaluate standardized and alternative sentences including fines, probation, incarceration, death, community service, electronic monitoring, etc.
18. Analyze the criminal justice system as it relates to politics, the media, and the community in general.
19. Identify employment opportunities and requirements in the criminal justice field.
20. Describe the types of terrorism and their impact on the U. S. and the role of the Department of Homeland Security.

## Course Format

The communication in this class will take place through lecture, discussion and various assignments that will be distributed by the instructor throughout the semester. The due dates of each assignment will be listed in the course schedule.

## Course Requirements

**COURSE ASSIGNMENTS:**
All assignments will have specific points assigned to them when handed out or assigned.


# GRADING

**Grades are calculated as follows:**

| Exams | Points |
|---|---|
| _____ | _____ |
| _____ | _____ |
| _____ | _____ |

## Assignments:

| | |
|---|---|
| Reflective Paper | _____ |
| Interview | _____ |
| Court Observation | _____ |
| Library Research Paper | _____ |
| Discussion Questions | _____ |
| Targeted Study Questions | _____ |
| **Total Possible Points:** | _____ |

If you fail to take an examination as scheduled the following will apply:
_____ you will receive a zero (0) score for the exam.
_____ **No makeup exams will be given.**

Exams will consist of a combination of Multiple Choice, True False, Quizzes or Short Answer Questions that will be taken at scheduled intervals. All tests are closed book. Tests will/will not be comprehensive examinations.

All Written Assignments will be evaluated for **completeness, accuracy, ability to follow requirements of the assignment, and integration of course material**.

The final course grade will be determined by the average of the student's scores on the all exams, and the assigned material based on a 100 point scale. **The reflective paper, interview and court observation, library research, targeted questions, and discussion questions are strongly considered as part of your participation grade.** It is important that all papers are turned in on the due dates. The instructor will strive to return papers to you within one week of the turn in date.

You are expected to participate in the course activities and discussions throughout the semester

## Increase Your Success

It is important to recognize that students will not be able to sufficiently learn criminal principles justice if you fail to read the course text, just provide unsupported conclusory generalizations in your written assignments, or try to get by with just reading the materials two hours prior to the scheduled exam dates. Do not wait until the last minute to begin or complete your assignments. Your inability to contact the instructor just hours before the assignment due date will not constitute a sufficient reason for delay in completing your required assignments.

Your constant diligence in reviewing the course materials on a weekly basis as scheduled, will ensure that you have a positive and successful learning experience in this course.

## Classroom or Online Conduct

Students should avoid any inappropriate personal comments or behavior directed toward fellow students or anyone else in during class activities or discussions. Any violations will be addressed when brought to the instructor's attention.

## Other Course Information

Meets College definition for:_____

This course is a _____elective and an _____ elective.

_____

Academic honesty, as defined in the Student Handbook, is required of all students.

# References

Alabama v. White, 496 U.S. 325, 110 S.Ct. 2412 (1990)

Article 27 Crimes and Punishment Maryland Code Ann. (1993)

Baltimore Examiner. *"10,000.00 fine imposed in sale of small crabs"*
    [by a Sun Reporter]. (October 4, 2007).

Bell, Chief Judge Robert M., *State of the Judiciary Address Before the
    Maryland General Assembly*. (February 13, 2003). Retrieved
    from the World Wide Web September 5, 2007: http://www.courts.
    state.md.us/soj2003.html

Broadwater, Luke, *Professor: Murder trial showed Columbia has a gang
    problem*. (January 31, 2007). Baltimore Examiner. Retrieved from
    the World Wide Web: October 13, 2007:
    http://www.examiner.com/a-538565~Professor_
    _Murder_trial_showed_Columbia_has_a_gang_problem.html

California law codes. Retrieved from the World Wide Web August 9, 2007:
    http://www.leginfo.ca.gov

Christo, Joe, *Citistat: One Way to Efficiently Run a City,* (2004). Center for
    Urban and Policy at Northwestern University, Boston, Spotlight.
    Retrieved from the World Wide Web September 23, 2007:
    http://www.curp.neu.edu/

City of Baltimore Website.  Retrieved from the World Wide Web
    July 3, 2007 http://www.ci.baltimore.md.us/news
    /citistat/index.html

Constitution of the United States

Corner, Candice. (September 1, 2006). *Ten Hot Jobs for 200.*:
    Retrieved from the World Wide Web September 1, 2006:
    http:// www.msn.com

District Court Charging Document Form DC/CR 44. ). Retrieved
    from the World Wide Web July 3, 2007:

http://www.courts.state.md.us/district/forms/criminal/dccr44.pdf

Dyson, Roy, Maryland Senator. *Legislation to Prevent Gang Activity Passes in Annapolis,* (May 6, 2007 ). Southern Maryland News Opinion. Retrieved from the World Wide Web October 7,2007: http://somd.com/news/headlines/2007/5872.shtml

Esebensen, Finn–Age, Tibbets, S. G. & Gaines, L. (2004). *American Youth Gangs at the Millennium.* Waveland Press, Inc., p.21.

Federal Bureau of Investigations Uniform Crime Reports Website. Retrieved from the World Wide Web: October 6, 2007: http://www.fbi.gov/ucr/03cius.htm

In re Appeal Misc. No. 32, 29 Md. App. 701 (1976)

In re Davis, 17 Md. App. 98 (1973)

In re Fletcher, 251 Md. 520 (1968),

In re Hamill, 10 Md. App. 586 (1970)

In re Leroy T., 285 Md. 508, 513(1979).

In re Victor B., 336 Md. 85 (1994)

In re Wooten, 13 Md. App. 521(1971)

Jackson v. State, 17 Md. App. 167 (1973)

John August Paulino v. State of Maryland, 399 Md. 341 (2007)

Kazeem Adeshina Ishola v. State of Maryland, 175 Md. App. 201 (2007)

Maryland Constitution Code Ann., Maryland Declaration of Rights, Article 5

Maryland Correctional Services Code Ann. §7-301 (2007)

Maryland Courts and Judicial Proceedings Code Ann. §3-8A-02 (2007)

Maryland Courts and Judicial Proceedings Code Ann. §3-8A-23 (2007)

Maryland Courts and Judicial Proceedings Code Ann. §4-301 (2007)

Maryland Courts and Judicial Proceedings Code Ann. §5-402 (2007)

Maryland Courts and Judicial Proceedings Code Ann. §8-108 (2007)

Maryland Criminal Law Code Ann. (2007)

Maryland Criminal Law Code Ann. § 2-202 (2007)

Maryland Criminal Law Code Ann. §10-114 (2007)

Maryland Criminal Procedure Code Ann. (2007)

Maryland Criminal Procedure Code Ann. § 10-105 (2007)

Maryland Criminal Procedure Code Ann. § 5-101

Maryland Criminal Procedure Code Ann. § 5-201 (2007)

Maryland Criminal Procedure Code Ann. § 6-216 (2007)

Maryland Department of Natural Resources Website Rules and Regulations.
    Retrieved from the World Wide Web: August 9, 2007:
    http://www.dnr.state.md.us/service/rulesandregs.asp

Maryland Department of Natural Resources Website News. Retrieved from
    the World Wide Web: September 23, 2007:
    http://www.dnr.state.md.us/dnrnews/index.asp

Maryland Emergency Management Agency Website. Retrieved
    from the World Wide Web: September 23, 2007:
    http://mema.state.md.us

Maryland Gang Information and Prevention Office Website. Retrieved
    from the World Wide Web: July 3, 2007:
    http://gangs.umd.edu/wfrmByLocation.aspx

Maryland General Assembly Website. *The Maryland Gang
    Prosecution Act of 2007, HB 713*. Retrieved
    from the World Wide Web: October 9, 2007:
    http://mlis.state.md.us/2007RS/billfile/hb0713.htm Bill file October 3

Maryland Governor's Office of Crime Control & Prevention Website.
    Retrieved from the World Wide Web: September 1, 2007:
    http://www.goccp.org/four/home.php

Maryland Judiciary Website, (2007). Retrieved from the World Wide
    Web: October 4, 2007:
    http://www.courts.state.md.us/coappeals/index.html and
    http://www.courts.state.md.us/coappeals/cert-petitions.html

Maryland Rules Code Ann. (2007)

Maryland Rules Code Ann. § 4-102(k) (2007)

Maryland Rules Code Ann. § 4-201 (2007)
Maryland Rules Code Ann. § 4-202 (2007)
Maryland Rules Code Ann. § 4-211 (2007)
Maryland Rules Code Ann. § 4-213 (2007)
Maryland Rules Code Ann. § 4-214(a) (2007)
Maryland Rules Code Ann. § 4-214(b) (2007)
Maryland Rules Code Ann. § 4-215 (2007)
Maryland Rules Code Ann. § 4-217 (2007)
Maryland Rules Code Ann. § 4-221 (2007)
Maryland Rules Code Ann. § 4-242 (2007)
Maryland Rules Code Ann. § 4-243 (2007)
Maryland Rules Code Ann. § 4-247 (2007)
Maryland Rules Code Ann. § 4-248 (2007)
Maryland Rules Code Ann. § 4-252 (2007)
Maryland Rules Code Ann. § 4-263 (2007)
Maryland Rules Code Ann. § 4-263 (2007)
Maryland Rules Code Ann. § 4-304 (2007)
Maryland Rules Code Ann. § 4-311(b) (2007)
Maryland Rules Code Ann. § 4-324 (2007)
Maryland Rules Code Ann. § 4-340 (2007)
Maryland Rules Code Ann. § 4-341 (2007)
Maryland Rules Code Ann. § 4-349 (2007)
Maryland Rules Code Ann. § 4-501 (2007)
Maryland Rules Code Ann. § 4-511 (2007)
Maryland Rules Code Ann. § 8-201 (2007)
Maryland State Bar Association Lawyers Manual (2007)
Maryland State Commission on Criminal Sentencing Policy. (2006)
    Retrieved from the World Wide Web: July 3, 2007:
    http://www.msccsp.org/guidelines/matrices.html#person and
    http://www.msccsp.org/publications/ar2006.pdf

Maryland State Police Recruiting Website. Retrieved from the World Wide
    Web: August 6, 2007: http://recruiting.mdsp.org/

Maryland Transportation Code Ann. § 26-202 (2007)

Maryland Uniform Citation *Advisement-Right to Counsel*. Retrieved
    from the World Wide Web: August 6, 2007:
    http://www.courts.state.md.us/district/forms/criminal/

McBrides, Don Home Page, *Large Cities With Highest Murder Rates*
    citing 2003 FBI - Uniform Crime Reports. Retrieved from the
    World Wide Web: September 23, 2007:
    http://www.geocities.com/dtmcbride/reference
    /murders_us_2003.html

Md. Criminal Law Code Ann. § 2-207 (a) (b) (2007)

Md. Criminal Law Code Ann. § 3-202 (2007)

Md. Criminal Law Code Ann. § 3-202 (2007)

Md. Criminal Law Code Ann. § 3-203 (2007)

Md. Criminal Law Code Ann. § 3-303 (2007)

Md. Criminal Law Code Ann. § 3-304 (2007)

Md. Criminal Law Code Ann. § 3-306 (2007)

Md. Criminal Law Code Ann. § 3-306(a)(2) (2007)

Md. Criminal Law Code Ann. § 3-306(a)(3) (2007)

Md. Criminal Law Code Ann. § 3-307(a)(3) (2007)

Md. Criminal Law Code Ann. § 3-307(a)(4) (2007)

Md. Criminal Law Code Ann. § 3-308 (a)(2)(3) (2007)

Md. Criminal Law Code Ann. § 3-308(a)(1) (2007)

Md. Criminal Law Code Ann. § 3-802 (2007)

Md. Criminal Law Code Ann. § 3-803 (2007)

Md. Criminal Law Code Ann. § 3-804 (2007)

Md. Criminal Law Code Ann. § 3-805 (2007)

Md. Criminal Law Code Ann. § 4-203 (2007)

Md. Criminal Law Code Ann. § 5-601(a) (2007)

Md. Criminal Law Code Ann. § 6-206 (2007)

Md. Criminal Law Code Ann. § 7-102(a)

Md. Criminal Law Code Ann. § 7-104 (a)(b)(c)(d) (2007)

Md. Criminal Law Code Ann. § 7-302 (2007)

Md. Criminal Law Code Ann. § 8-301( c)

Md. Criminal Law Code Ann. § 10-114 (2007)

Md. Criminal Law Code Ann. § 10-115 (2007)

Md. Criminal Procedure Code Ann. § 4-101 (2007)

Md. Transportation Code Ann. § 26-203 (2007)

O'Connel, Paul E.. (August 2001) *Using Performance Data for Accountability*. Pp24 -27. Retrieved from the World Wide Web: September 23, 2007: http://www.businessofgovernment. org/pdfs/Oconnell_Report.pdf

O'Guinn, Patrick J., Sr. & Nithianandam,V. (2007). Computer Forensics 101 Course, Howard Community College

Office of Postsecondary Education of the U. S. Department of Education Website,Campus Security Data Analysis Cutting Tool Website. Retrieved from the World Wide Web: September 27, 2007: http://ope.ed.gov/security/\

Prout v. State, 311 Md. 348 (1988)

Quickfacts, U.S. Census Maryland Population Statistics by Race. Retrieved from the World Wide Web: October 4, 2007: http://quickfacts.census.gov/qfd/states/24000.html

Security on Campus, Inc. College and University Campus Crime Statistics. Retrieved from the World Wide Web: September 27, 2007: http://www.securityoncampus.org/crimestats/index.html

State v. Earmon Alvin Wallace, Sr., 142 Md. App. 673, cert granted, 369 Md. 301(2002)

Stevenson and Wilson v. State, 287 Md. 504 (1980)

The Constitution of Maryland (Art. I, § 12; Art. IV, § 2). Retrieved from

the World Wide Web: September 5, 2007:
http://www.courts.state.md.us/judgeselect/judicialvacancy.html
Timms v. State, 83 Md.App. 12, 16-19 (1990)
U.S. Attorneys Office Website: Maryland Exile. Retrieved from
the World Wide Web: July 3, 2007:
http://www.usdoj.gov/usao/md/Exile/files/Baltimore%20EXILE%
20 strategy%20final.2006.02.15.pdf
U.S. Census Bureau, ePodunk Population Growth of Counties 2003.
Retrieved from the World Wide Web: August 8, 2007:
http://www.epodunk.com/top10/countyPop/coPop21.html
Webster,Merriam Dictionary Online. Retrieved from
the World Wide Web: October 4, 2007:
http://www.m-w.com/cgi-bin/dictionary?forensic
Westcott v. State , 11 Md. App. 305, cert denied, 262 Md. 750 (1971)